DEFINITE

A System to Support DEcisions on a FINITE Set of alternatives

User Manual

Environment & Management

VOLUME 3

The titles published in this series are listed at the end of this volume.

DEFINITE

A System to Support DEcisions on a FINITE Set of alternatives

User Manual

by

Ron Janssen
and
Marjan van Herwijnen

Free University Amsterdam,
Institute for Environmental Studies,
Amsterdam, The Netherlands

SPRINGER-SCIENCE+BUSINESS MEDIA, B.V.

A C.I.P. Catalogue record for this book is available from the Library of Congress

ISBN 978-94-010-4361-8 ISBN 978-94-011-0838-6 (eBook)
DOI 10.1007/978-94-011-0838-6

Printed on acid-free paper

Table of Contents

ACKNOWLEDGEMENTS

DEFINITE has been developed at the Institute for Environmental Studies of the Free University of Amsterdam. The Dutch version of the program was commissioned and developed in cooperation with the Ministry of Finance of the Netherlands. Without their support this program would not have been made. The procedures for sensitivity analysis within DEFINITE were developed in close cooperation with Piet Rietveld.

Many people have helped in preparing DEFINITE for publication. Among these people we would like to thank Ian Priestnall (Paragraph Services) for editing the program and user manual, Age Akkerman for typing the user manual, Els Hunfeld for her finishing touch to the user manual, Robbert van der Hoorn (GRIP Automation) for testing the program and David James (Ecoservices) for his comments. A special word of thanks to Henny Hoogervorst for her continuous and cheerful efforts to get this complicated project on its way. Many thanks also to Wim Hafkamp for his contribution to getting DEFINITE started and to Harmen Verbruggen for his endurance in getting it completed.

Part 1

Getting started

1. SETTING UP DEFINITE

1.1. What is DEFINITE?

DEFINITE (decisions on a finite set of alternatives) is a decision support software package that has been developed to improve the quality of environmental decision making. DEFINITE is, in fact, a whole toolkit of methods that can be used on a wide variety of problems. If you have a problem to solve, and you can identify alternative solutions, then DEFINITE can weigh up the alternatives for you and assess the most reasonable alternative. The program contains a number of methods for supporting problem definition as well as graphical methods to support representation. To be able to deal with all types of information DEFINITE includes five different multicriteria methods, as well as Cost-Benefit and Cost-Effectiveness analysis. Related procedures such as weight assessment, standardization, discounting and a large variety of methods for sensitivity analysis are also available.

DEFINITE supports the whole decision process, from problem definition to report generation. The structured approach adopted ensures that the decisions arrived at are systematic and consistent.

DEFINITE can be used by the busy professional with no prior experience of such software, as well as the sophisticated user. A tutorial example and examples from the practice of environmental decision making are provided. Menus, information screens and help screens will lead you through the program and will very rapidly make you familiar with its features.

The DEFINITE package

The DEFINITE package comprises:

- the book *Multiobjective Decision Support for Environmental Management*;
- the DEFINITE program disks;
- the DEFINITE reference manual and practical examples;
- a help desk for user support.

A short description of the program can be found in Chapter 5 of *Multiobjective Decision Support for Environmental Management*. Detailed technical descriptions of all methods available in DEFINITE are included in Chapters 3 and 4. The DEFINITE program disks contain the program. This chapter of the manual will help you setting up DEFINITE. First we will tell you how to reach the help desk (Section 1.2) and provide you with a specification of the equipment you need to work with DEFINITE (Section 1.3). This is followed by instructions on installation (Section 1.4) and information on the use of the three parts of the manual (Section 1.5).

Conventions

In this manual bold print is used to highlight input from the keyboard. A distinction is made between **bold print**, **BOLD PRINT** and **<BOLD PRINT>**.

bold print refers to options as they appear in a menu or command line. You can, for example, select **create an effects table** from the menu of module A and select **help** from the command line of this procedure.

BOLD PRINT refers to input you need to enter from the key board. For example, in Part 2 you are instructed to type: **DEMO** to name your session.

<BOLD PRINT> refers to specific keys on the keyboard.
<F1> refers to function key 1. Function keys can be used to make a selection from a menu or command line.
<SPACE> refers to the space bar.
<RETURN> refers to the return or enter key. You can use **<SPACE>** to move through a menu followed by **<RETURN>** to select the highlighted option.
The **<CURSOR>** or **<ARROW>** keys are used to move the cursor within a table.
The **<ESC>** can be used to cancel a procedure and return to the previous step.

1.2 The help desk

Should you have any questions or problems using this software, please fax the help desk for assistance. The fax number for the help desk nearest you is listed on the back of the slipcase which accompanies this manual.

When faxing the help desk, clearly state the following:

- your fax number;
- the type & set-up of the computer you are using;
- the procedure you were using when you encountered the problem;
- the last action before you encountered the problem; and
- a description of the problem.

We are always happy to be of assistance and we will endeavor to fax a reply to you as soon as possible.

1.3. Equipment you need

To be able to run DEFINITE you need the following equipment:

- an IBM XT or AT personal computer or 100% compatible;
- a 3½ inch 720 KB diskdrive;
- a hard disk with at least 2.5 MB available;
- 640 KB memory with at least 570 KB available;
- an EGA, VGA, or CGA graphics card;
- MS-DOS 2.0 or higher.

DEFINITE uses 570 KB memory. If you have resident programs on your computer that require a lot of memory space it is better to deactivate them while running DEFINITE. If you do not know how much memory you have available just run the program. DEFINITE provides a message if the necessary memory is not available. Free memory and available space on the hard disk is reported by the MS-DOS **CHKDSK** or **MEM** command.

A special 3½ inch to 5¼ inch copy utility is available for users who do not have a 3½ inch diskdrive but have access to a computer with both types of diskdrives. Type **COPY360K <RETURN>** to run this utility from disk 1. Make sure you label the 5¼ inch disks immediately as disk 1, disk 2 etc.. The new disk 1 includes the Install program for 5¼ inch disks.

1.4. Installing DEFINITE

To be able to run DEFINITE from the hard disk you have to do the following to install the program on hard disk:

1. Make sure the DEFINITE disks are in the read only position
2. Insert the DEFINITE disk '1:3' into disk drive A
3. Create a DEFINITE directory on the hard disk: **MD DEFINITE <RETURN>**
4. Change to the DEFINITE directory: **CD DEFINITE <RETURN>**
5. Start the installation program: **A:INSTALL <RETURN>**
6. Follow the instructions or the screen and insert disks '2:3' and '3:3' when asked.

1.5. About this manual

This manual consists of three parts.

Part 1. **Getting started**: helps you set up DEFINITE and includes a tutorial to give you an idea of the potential of DEFINITE (Chapter 2).

Part 2. **Reference Manual**: provides you with detailed descriptions of all procedures. Each description includes a short summary of the procedure, the required input, the type of output, an example and information on default settings of a procedure. These descriptions are also available if you select **help** from within the program.

Part 3. **Examples from the practice of environmental decision making**: gives you a guided tour through four practical examples.

The manual does not include descriptions of the methods used in DEFINITE such as multicriteria methods, Cost-Benefit analysis and methods for sensitivity analysis. These descriptions are available in Chapters 3 and 4 of *Multiobjective Decision Support for Environmental Management*. Technical details of the program can be found in the Appendix.

2. TUTORIAL

2.1. The Nyverhoeck example

A simple example is used to illustrate the various methods. This example is also used in the book. We recommend that you have a look at Chapter 5 of *Multiobjective Decision Support for Environmental Management* before or while you go through the example.

Figure 2.1 shows a map of an imaginary region in the south-west of the Netherlands. The road through the centre of the region connecting the north with the industrial port of Nyverhoeck has insufficient capacity; as a result frequent traffic jams occur. To improve this situation three alternatives are suggested:

1. A four-lane highway to the west and parallel to the existing road, following the coastline;
2. The same but divided into two lanes for private and two lanes for public transport;
3. A train link entering the region from the north-west, passing through the middle of the region, and ending at Nyverhoeck.

2.2. Start the example

Start DEFINITE: **DEFINITE <RETURN>**
Read the introductory screens, hitting any key to continue: **<ANY KEY>**
Create a session to work with. Name this session demo: **TUTORIAL<RETURN>**
Answer yes to the question: Do you want to use the results of a previous session?: **Y**
Move the cursor to EXAMPLE and select the **example**: **<CURSOR><RETURN>**

2.3. Problem definition

Start **module A: problem definition**: **<F1>**
Leave the information screen: **<ANY KEY>**

Information screens are presented upon entering each procedure. You can leave each information screen by hitting **<ANY KEY>**. This instruction is not repeated in the remainder of the example.

Start **create an effects table**: **<F1>**
The other procedures available for problem definition will be shown at the end of the example.

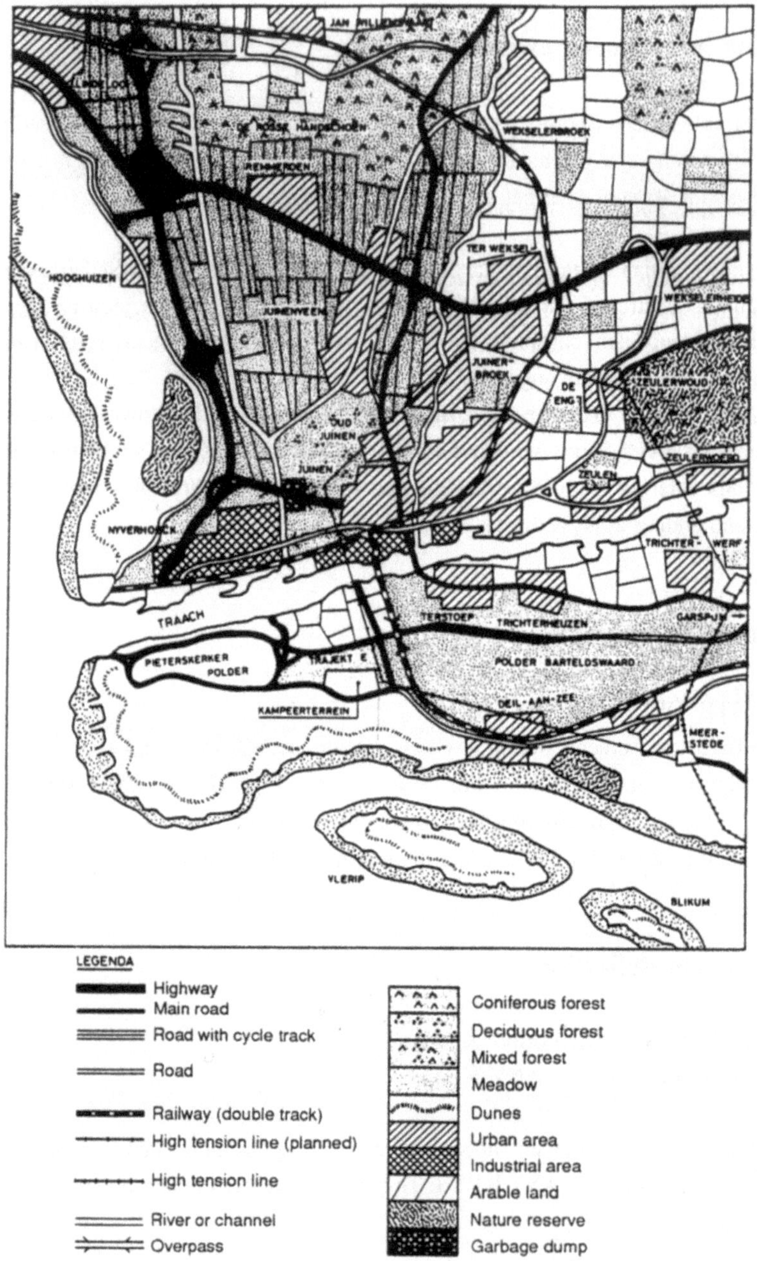

Figure 2.1. Map of the plan region.

After the information screen you will now see the effects table of the example (Screen 2.1). DEFINITE recognizes costs (the lower the better) by their negative sign. In this table Emissions NO_x and Costs are costs and have therefore received a negative sign.

Screen 2.1. Effects table. Improve transportation to Nyverhoeck.

```
effects table       Highway  Road/bus Train
===========================================================================

===========================================================================
Costs                -200.00  -250.00  -400.00
Travel time              +++       ++        +
Capacity               20.00    30.00    40.00
Emissions NOx       -1000.00  -750.00  -100.00
Landscape                ---      ---        -






Select an option:
==Add an alternative=============================================A1   ==
F1.alternative  F2.effect        F3.descriptor    F4.edit         F5.units
F6.time         F7.information   F8.help          F9.continue     F10.more
```

Use the space bar to obtain information on the various options: **<SPACE>**
Leave this procedure: **<F9>**

If you have made any alterations to the effects table DEFINITE will ask you whether you want to save these changes. Answer no to this question: **N**
Continue with **module B: problem presentation**: **<F2>**

2.4. Problem presentation

Review the description of all methods available in DEFINITE: **<F1>**
Select **stop** to return to the menu and and then select **graphic presentation**: **<F3><F3>**
Rank the effects in descending order of priority by exchanging rows: **<F1>**
Rank the alternatives: **<F4>**

Screen 2.2. A graphical presentation of the effects table.

Improve transportation to Nyverhoeck.

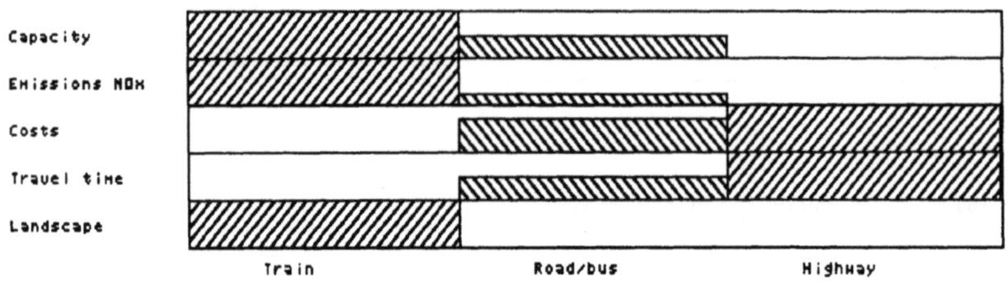


```
Select an option:
=Rank the alternatives ================================================
F1.exchange row F2.exchange clm F3.print        F4.rank       F5.information
F6.Help         F7.continue
```

Leave this procedure and continue to evaluation: **<F7><F5>**

2.5. Problem evaluation

Select **multicriteria analysis**: **<F1>**
Select the **Expected Value method**: **<F5>**

To be able to apply any multicriteria method you need to assign weights (priorities) to the effects. Select **weights**: **<F2>**
Select the **Expected Value method**: **<F1>**
Select **make**: **<F1>**
Select **Capacity** as the most important effect: **<SPACE><SPACE><RETURN>**
Select **Emissions** NO_x as the next effect: **<SPACE><SPACE><RETURN>**
and make it equally as important as the previous effect: **<F2>**
Continue with the remaining effects and hit escape to confirm the weights: **<ESC>**
If you like you can add a line of information to this set of weights
Answer no to the question "You want to replace the last Expected Value weights?": **N**

Now that you have assigned weights you can execute the Expected Value method: **<F1>**
First DEFINITE shows you the standardized scores: **<ANY KEY>**
The results of the Expected Value method are now shown (Screen 2.3). In the left column you see the weights and in the right column the scores and ranking of the alternatives.

Screen 2.3. Results of the Expected Value method.

Improve transportation to Nyverhoeck.

```
═══════════════════════════Expected Value Meth.═══════════════════════════

-----------WEIGHTS---------               ----------RANKING--------
RANKING                  SCORES           RANK                SCORES
 1: Capacity              0.357            1: Train            0.82
    Emissions NOx         0.357            2: Road/bus         0.50
 3: Costs                 0.157            3: Highway          0.27
 4: Travel time           0.065
    Landscape             0.065



Select an option:
══Execute this multicriteria method═════════════════════════════════C2.4══
F1.execute       F2.weights      F3.standardize  F4.save results F5.print result
F6.effects tab.  F7.graph        F8.other MCA    F9.help         F10.continue
```

Select **other MCA** and **Regime method** to switch to this method: **<F8><F4>**
Since you have already defined weights you can execute the method directly: **<F1>**
First you obtain a probability table. Each cell entry in this table is the probability that the alternative indicated by the row is ranked ahead of the alternative indicated by the column. Continue to obtain the ranking of the alternatives: **<F1>**

The results of the Regime method are now shown (Screen 2.4). In the left column you see the weights and in the right column the scores and ranking of the alternatives. Note that only the ordering of the effects is used; this method does not require quantitative weights. Since a random generator is used to produce the results small differences may occur each time the method is applied.

Screen 2.4. Results of the Regime method. Improve transportation to Nyverhoeck.

```
==============================Regime Method==============================

-----------WEIGHTS---------               ----------RANKING--------
RANKING                                   RANK                SCORES
 1: Capacity                               1: Train             1.00
    Emissions NOx                          2: Road/bus          0.45
 3: Costs                                  3: Highway           0.05
 4: Travel time
    Landscape




Select an option:
=Execute this multicriteria method===================================C2.3==
F1.execute      F2.weights      F3.standardize  F4.save results F5.print result
F6.effects tab. F7.graph        F8.other MCA    F9.help         F10.continue
```

If you like you can generate a graph of these results: **<F7>**

Answer no to the question "Do you want to change information?": **N**

Do not replace the existing results of the Regime method: **N**

Do not save any results and continue to **module D: sensitivity analysis**: **<F4><F10>Y<F2>**

2.6. Sensitivity analysis

Select **method uncertainty**: **<F1>**

Select the weights used in the multicriteria methods (**wexp.001**): **<RETURN>**

(Select **regime.001** if that choice is offered)

DEFINITE tells you which results have been saved in this example.

Select **calculate** to obtain the results **<RETURN>**

Select **Line graph** and the number of alternatives to be compared: **<F2>3<RETURN>**

Select **Highway**, **Road/bus** and **Train**: **<F1><F2><F3>**

This graph shows the results of the various methods applied to the same effects table (Screen 2.5). As you can see in this example all multicriteria methods produce the same ranking of alternatives.

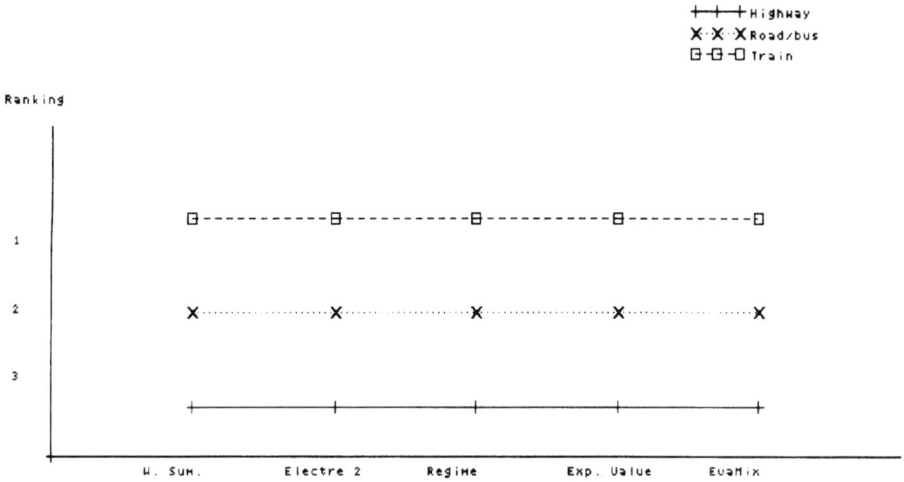

Select an option:
=Print the diagram
F1.print F2.continue

Leave the graph and ask for a conclusion: **<F2><F1>**

Select **other analysis** to leave this procedure and choose **weight intervals**: **<F8><F4>**
Select the results of the Expected Value method **(exp.001)**: **<ANY KEY><F4>**
Select interval and the alternatives **Train** and **Highway** for comparison: **<F1><F3><F1>**
Select the weight of **Travel time** to be analyzed: **<F2>**
DEFINITE now produces weight intervals for travel time (Screen 2.6).

The effect intervals procedure produces the same type of results for effect scores. Select **other analysis** followed by **effect intervals** to try this procedure: **<F8><F5>**

Select **other analysis** followed by **effects uncertainty**: **<F8><F2>**
This procedure tests the sensitivity of the ranking to uncertainties in the effect scores.
Select the results of the Expected Value method to be analyzed: **<F4>**
Select **calculate** and assign an uncertainty of 25% to all scores: **<F1>25<RETURN>**

DEFINITE now produces a probability table (Screen 2.7). This table shows the probability that an alternative, given the uncertainty in the scores, is the best, second best or worst alternative.

Screen 2.6. Weight interval. Improve transportation to Nyverhoeck.

```
═══════════════════════════ WEIGHT INTERVAL ═══════════════════════════

      1 ------
        ------                          MCA method       : Expected Value Meth.
        ------                          Weight method    : Expected Value
        ------
        ------ Highway  > Train
        ------
        ------
        reverse   0.608                 The weight interval is calculated
        ++++++                          for effect        : Travel time
        ++++++                          with rank number :   4
        ++++++                          and weight        :   0.065
        ++++++
        ++++++                          A rank reversal takes place
        ++++++                          at value :     0.608
        ++++++
  0.065 start  Train     > Highway
        ++++++
      0 ++++++

Select an option:
═Save this interval═══════════════════════════════════════════════D4 ══
F1.save        F2.print       F3.information  F4.help        F5.other weight
F6.other alter. F7.other MCA  F8.other anal.  F9.continue
```

Screen 2.7. Results of effects uncertainty. Improve transportation to Nyverhoeck.

```
probabilities      Train     Road/bus Highway
════════════════════════════════════════════════════════════════════


════════════════════════════════════════════════════════════════════
rank  1                0.99      0.01
rank  2                0.01      0.99
rank  3                                    1.00
total score            2.99      2.01      1.00





Select an option:
═Display conclusion═══════════════════════════════════════════════D2 ══
F1.conclusion   F2.save         F3.print        F4.percentages  F5.information
F6.help         F7.other perc.  F8.other mca    F9.other anal.  F10.continue
```

Select **conclusion** for an interpretation of the table: **<F1>**
Leave this procedure, return to the **main menu**: **<ANY KEY><F10><F2>**
You can now **exit the program**: **<F7>**
DEFINITE includes similar procedures to test the sensitivity of results of Cost-Benefit analysis.

2.7. Other procedures to support problem definition

DEFINITE includes other procedures to support problem definition. We show below how to use the **combine alternatives** procedure.

Start **problem definition** from the main menu: **<F1>**
Start **combine alternatives**: **<F4>**

To secure the water supply for Deil aan Zee (see Figure 2.1) a minimum of 10,000 m^3/year needs to be extracted from groundwater stocks. Four wells are available in the dune area nearby. These wells (shown as elements 1 to 4 in the table) have a capacity of 3000, 5000, 2000 and 6000 m^3 year. Continue to the combination rule: **<F9>**

The combination rule specifies that each alternative must produce at least 10,000 m^3/year and no more than 12,000 m^3/year. Continue to generate the alternatives: **<F9>**
DEFINITE now generates all alternatives that comply with the combination rules (Screen 2.8). Select **composition** to see how the elements are combined into alternatives: **<F3>**

Screen 2.8. Effects table. Water supply for Deil aan Zee.

Combined	Alt.01	Alt.02	Alt.03
Capacity	1.0E+04	1.1E+04	1.1E+04
Land use	-800.00	-1100.00	-1100.00
Groundwater leve	-19.00	-24.00	-25.00

```
Select an option:
==Edit column headings================================================A4 ==
F1.edit          F2.units          F3.composition  F4.print          F5.information
F6.help          F7.continue
```

Return to combination rules to edit the required capacity: **<F1><ESC><F3>**
Select **Capacity: <RETURN>**
Enter a new minimum and maximum value: **9000<RETURN>12000<RETURN>**
Select **continue** to combine the elements into 4 alternatives: **<F9>**
Select **continue** to leave this procedure and save the results: **<F7> Y**
If you save this table you create an effects table. Since, in any single session, only one effects table can be consistent with the evaluation results, the previous results will be deleted as well as the previous effects table. You could now rerun the whole example using this new effects table.

You now enter procedure A5: **discard irrelevant alternatives: <F4>**
If you select **inefficient** you will see that all alternatives are dominated by alternative 1: **<F2>**

Other procedures to support problem definition are a procedure to generate a list of effects and a procedure to generate a list of alternatives. After you have entered one of these procedures select **help** for an explanation of the procedure.

2.8. How to restore the original example

If you made changes to the example you might want to restore the original example. This can be done by leaving the program and starting DEFINITE as shown above (Section 2.2).

The session to work with is still called TUTORIAL; type the name: **TUTORIAL<RETURN>**
Answer no to the question "Do you want to continue with the results of this session?": **N**
Answer yes to the question "Do you want to use the results of a previous session?": **Y**
Select the example: **<CURSOR><RETURN>**

DEFINITE now copies the EXAMPLE session to the TUTORIAL session. You will work in the TUTORIAL session and the example will remain unaltered and can be used again later. The example is now ready to start. The original example can also be restored by selecting **utilities** from the main menu followed by **change**: **<F6><F1>**

2.9. How to create your own example

We suggest that you now continue with one or more of the four available practical examples that are described in Part 3 of this manual. If you feel confident enough with the program you can also create your own example to experiment with.

Go to Part 3 to continue with one of the practical examples.

Create a new session to be able to enter your own example.
Enter **utilities** from the main menu: **<F6>**
Change to a new session: **<F1>**
Enter a name for your session: **YOURNAME<RETURN>**
Answer no to the question "Do you want to use the results of a previous session?": **N**

Return to the main menu: **<F8>**
Start **problem definition** from the main menu: **<F1>**
Start **create an effects table**: **<F1>**
Enter your **alternatives**: **<F1>**
Enter your **effects**: **<F2>**
The program now asks you for the measurement scale of the effects you enter. Select **help** if you are not familiar with these scales: **<F8>**
Continue: **<F9>**

DEFINITE will now ask you if you want to save this table. We suggest that you use your own example to experiment further with DEFINITE. Maybe you can explore the facilities for **Cost-Benefit analysis** or experiment with the possibilities of the appraisal table. Using the **appraisal table** procedure you can generate the diagrams as shown in Chapter 7 of *Multiobjective Decision Support for Environmental Management*. To get an appraisal table you need to define two or more appraisal categories and calculate the appraisal table. Select **diagram** to generate the diagrams. Select **help** for information on the various available options. Select **utilities** followed by **information** if you want to skip the information screens.

2.10. What next?

You now have an idea of the potential of DEFINITE. To see how DEFINITE works in practice and to have more procedures demonstrated you could continue with the examples in Part 3. Detailed descriptions of all procedures are available in Part 2. These descriptions are also available if you select **help** from within the program.

Part 2

Reference manual

3. PROCEDURES OF THE MAIN MENU AND UTILITIES

Each problem is stored in a separate session. As a first step you need to specify the name of your session. Next you need to indicate whether you want to enter a new problem in this session or whether you want to continue with the results stored previously.

DEFINITE includes 5 modules:

- A: problem definition
- B: problem presentation
- C: problem evaluation
- D: sensitivity analysis
- E: report generation

You enter the five modules in DEFINITE through the main menu. A flow diagram of the DEFINITE modules is shown in Figure 3.1. A flow diagram of each of the five modules and detailed descriptions of the procedures available in these modules are included in the Chapters 4 to 8.

The main menu also provides access to a number of utilities. Utilities can be used to:

- start a new session;
- switch the presentation of intermediate results on or off;
- make a backup of a session on diskette;
- switch the information screens presented upon entering each procedure on or off;
- change the setting of the appraisal table.

DEFINITE menus

Upon entering the module, you will see the module menu. This menu provides access to the procedures available in this module. A procedure is executed through its command line at the bottom of the screen.

There are two ways to make a selection from a menu:

1. Press **<SPACE>** to move through the menu and hit **<RETURN>** to select the highlighted option. An information line above the command line provides information on each highlighted option.
2. Hit the function key that corresponds to the option you want to select (e.g. **<F1>** to select the first option in a menu).

In every command line you can:

- Select **continue** to complete a procedure and to continue to the next step in the program;
- Hit **<ESC>** to cancel a procedure and return to the previous step;
- The **<CURSOR>** or **<ARROW>** keys are used to move the cursor within a table.

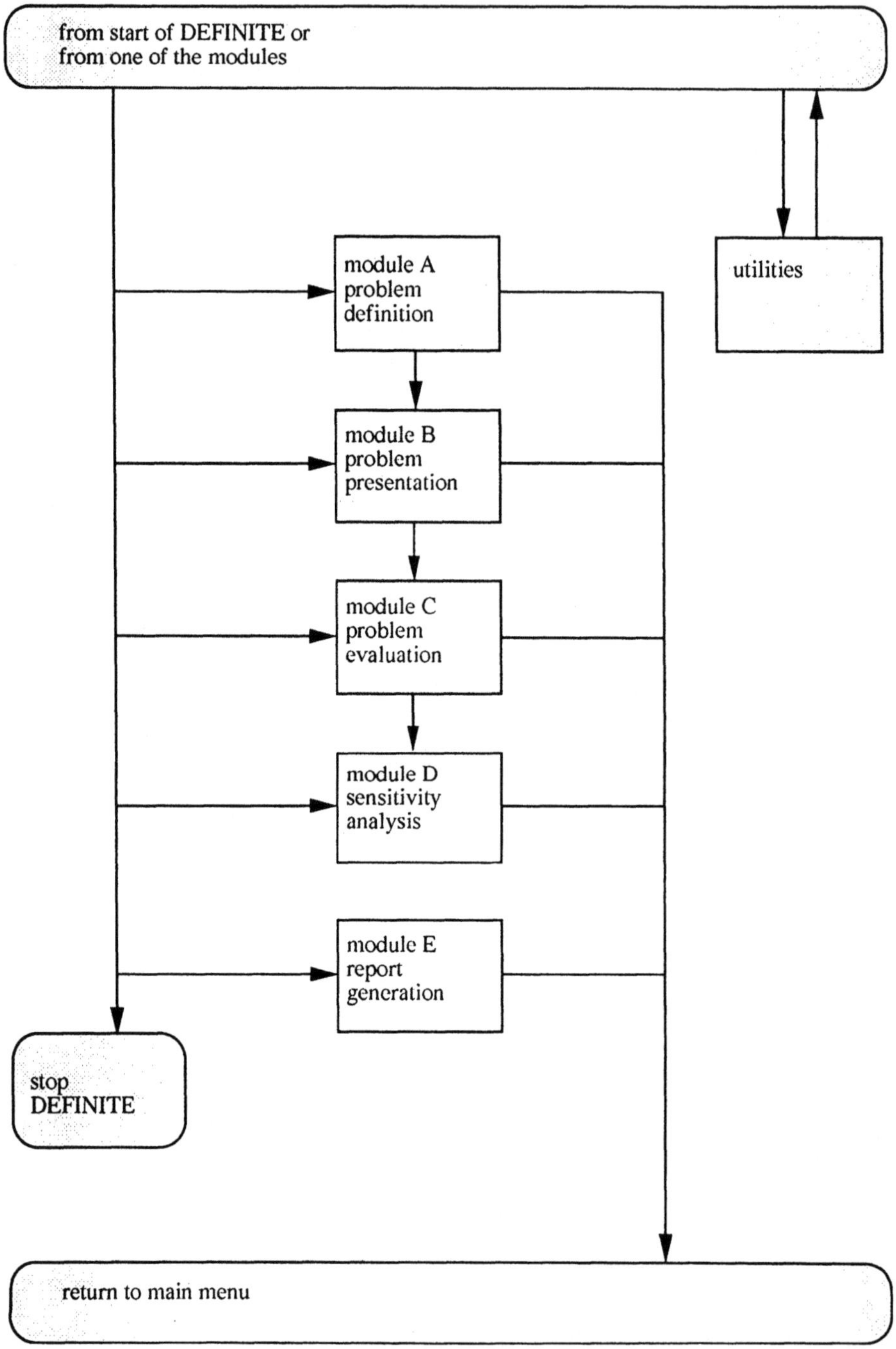

Figure 3.1. Main menu and utilities.

4. PROBLEM DEFINITION PROCEDURES

The procedures available in **module A: problem definition** can be used to generate an effects table that is as complete as possible. Problem definition is supported by the following procedures.

A1. **Create an effects table** to enter the effects table directly using a spreadsheet environment.

A2. **Generate a list of effects** if support is required to identify effects to be included in the effects table.

A3. **Generate alternatives** if it is possible to 1: distinguish various aspects or dimensions of the problem, and 2: identify per aspect one or more solutions.

A4. **Combine alternatives** if it is possible to combine elements or parts of alternatives into alternatives.

A5. **Discard irrelevant alternatives** to eliminate alternatives that are not relevant to the decision problem.

From procedure A1 you can enter the procedures A1.1 and A1.2.

A1.1. **Discount future effects** to discount effects tables for future years to a discounted effects table for a common base year.

A1.2. **Assess scores** can be used if no quantitative estimates are available for the scores of all alternatives on one effect. An estimate of the relative size of the effect for each alternative is provided using pairwise comparison of the alternatives.

A flow diagram of the procedures of **module A: problem definition** is shown in Figure 4.1.

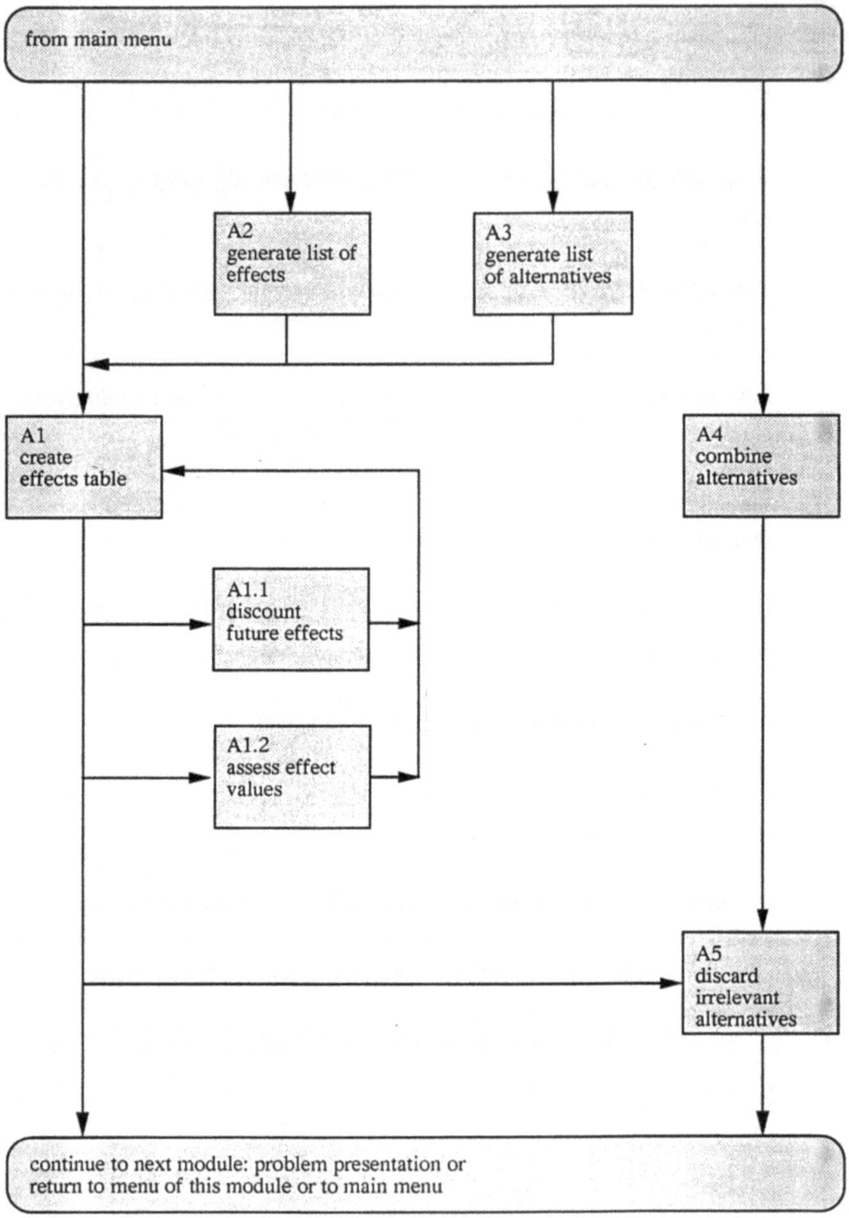

Figure 4.1. Procedures module A: problem definition.

DEFINITE menus

Upon entering the module, you will see the module menu. This menu provides access to the procedures available in this module. Each procedure is executed through its command line at the bottom of the screen.

There are two ways to make a selection from a menu:

1. Press the **<SPACE>** bar to move through the menu and hit **<RETURN>** to select the highlighted option. An information line above the command line provides information on each highlighted option.
2. Hit the function key that corresponds to the option you want to select (e.g. **<F1>** to select the first option in a menu).

In every command line you can:

- Select **continue** to complete a procedure and to continue to the next step in the program;
- Hit **<ESC>** to cancel a procedure and return to the previous step;
- The **<CURSOR>** or **<ARROW>** keys are used to move the cursor within a table.

PROCEDURE A1. CREATE AN EFFECTS TABLE

Purpose of this procedure

This procedure creates an effects table. An effects table includes the names of the alternatives (the column headings) and the names of the effects (the row headings). The cell values of the table represent the size of the effects for each alternative. The effects table is the central element of a session. A session can include only one effects table. If you want to use two or more effects tables you need to create a separate session for each effects table. Existing effects tables may also be modified with this procedure.

Your input and DEFINITE's output

To create the effects table enter:

1. Names of the **alternatives**. You can enter a long name (16 characters) and a short name (8 characters); the short name serves as a column heading.
2. **Effects**. For each effect you need to enter:
   - the name of the effect;
   - the measurement scale, e.g. **ratio** scale or **---/+++** scale;
   - for quantitative scales: the measurement unit, e.g. km/hour.
   - **benefit**, if a higher score is better than a lower, e.g. income; or **cost**, if a lower score is better than a higher, e.g. emissions.
3. **Effect scores.**

This results in:

A completed effects table.

Example

The road from the north to the industrial port of Nyverhoeck has insufficient capacity. Three alternatives are available to solve this problem.

Your input

Enter:

1. **Alternatives.** In this example Highway, Road/bus, Train.
2. **Effects.** Costs in mln $, reduced Travel time in minutes per person per day, Capacity in mln kilometres per year, Emissions NO_x in tons per year and, finally, loss of Landscape in hectares. Costs, Emissions NO_x and Landscape are costs (the lower the better); Capacity and Travel time are benefits (the higher the better).

3. **Effect scores.** For example, Costs of 200 mln $ for the Highway alternative.

DEFINITE's output

Effects table. Improve transportation to Nyverhoeck.

effects	units	Highway	Road/bus	Train
Costs	mln $	-200	-250	-400
Travel time	min/prs/day	30	15	10
Capacity	mln/km/yr	20	30	40
Emissions NO_x	ton/yr	-1000	-750	-100
Landscape	hectare	-10	-12	-2

Note that Travel time reflects the reduction in travel time (benefit) and Landscape reflects the loss of landscape (cost) that result from the alternatives.

Effects defined as costs (Costs, Emissions NO_x and Landscape) have received a minus sign.

Comments

Dimensions of the effects table

An effects table can include a maximum of 50 alternatives, 50 effects and 5 descriptors.

Alternatives

To each alternative you can give a long name (at most 16 characters) and a short name (at most 8 characters). The short name appears in the table, the long name is displayed with the **units** option. The long name is printed in combination with the effects table and is included in the report. If the long name does not exceed 8 characters, the long name and short name are identical. The cursor can be moved to the top line of the table using **<PgUp>** to edit the names of the alternatives.

Definition of effects

Effects described as costs, and which are measured using a ratio, interval or monetary scale, will automatically receive a minus sign. Effects measured on an ordinal scale will obtain a rank number; on an ordinal scale a rank of 1 represents the best score. The scores +, ++ and +++ indicate that the alternatives differ, but that it is not clear to what degree. Units are shown after pressing function key **<F5>** (units).

Effects defined as benefits (the higher the better) are displayed as positive numbers, effects defined as costs (the lower the better) are displayed as negative numbers. It is not possible to enter a negative score for an effect defined as a benefit or a positive score for an effect defined as a cost.

It is possible to change the measurement scale and unit of an effect or to change cost into benefit or vice versa. To do this you need to select **units** followed by **change**. To monetary effects you can assign x1, x1000, millions and billions. The last monetary unit you define determines the unit for all monetary units.

Measurement scales

The effects of the alternatives can be entered into the effects table in qualitative and quantitative form. The following measurement scales are available: ratio scale, interval scale, monetary scale, ordinal scale, ---/+++ scale, binary scale and nominal scale. The measurement scale plays an important role in the evaluation of the effects table. A short summary of the characteristics of the available measurement scale is provided below.

The importance of an effect measured on a **ratio scale** is proportional to its value. Examples are the inhabitants of a city or the speed of a car.

The **monetary scale** is a special case of the ratio scale. Effects on a monetary scale are measured in monetary units such as dollars, guilders etc.. Examples are income generated by a project or the investment costs of a project.

An **interval scale** does not have a meaningful origin. This implies that only differences between effects are meaningful. An example is the temperature as measured in degrees Celsius. The statement that 40 degrees Celsius is twice as warm as 20 degrees Celsius is incorrect. However, the statement that the difference between 20 and 40 degrees Celsius is two times the difference between 10 and 20 degrees Celsius is correct.

Effects measured on an **ordinal scale** are ranked only. For example the visual effect on the visual quality of the landscape of various alternatives can be expressed by indicating which alternative has the least effect, which alternative is second best, until the worst alternative is reached. No statements can be made with regard to the size of the differences between alternatives.

The **---/+++** scale is a special case of the ordinal scale. The effects are ranked by assigning one or more pluses or minuses.

A **binary scale** only indicates whether an effect occurs or does not occur. It is necessary to indicate whether 'yes' should be considered better than a 'no' or vice versa. Violation or non-violation of an environmental standard is an example.

Effects measured on a **nominal scale** cannot be used to rank the alternatives. These effects are no more than a specification of an alternative. Examples are the name of a location or the colour of an alternative.

Display and save effects

An effect is shown in the effects table with a maximum of 4 digits before the decimal point and 2 digits after the decimal point. Large or small numbers are displayed in exponential notation. For example 5200000 is displayed as 5.2E+06 and 0.0000052 as 5.2E-06. Scores can also be entered in exponential notation.

Irrespective of the position of the decimal point after rounding, at most 6 digits are saved on disk. Within each module all calculations are performed with a minimum of 4 decimals.[1] The following examples show the implications for accuracy.

You enter	Screen display	Disk storage
12345678	1.2E+07	1.23457E+07
0.12345678	0.12	1.23457E-01
0.00000012345678	1.2E-07	1.23457E-07

Descriptors

Descriptors can be added to further characterize or describe an alternative. Examples are the size, colour or location of an alternative. Descriptors play no role in the evaluation of alternatives.

Changing an existing effects table

All tables and results produced in modules B, C, D and E are linked to the effects table produced in module A. To guarantee consistency within one session a change in the effects table results in the deletion of all previous results. A warning is provided before the effects table is saved.

An exception to this rule applies if the charges in the effects table are limited to: 1. effect scores, and 2. number or names of alternatives. In this case a definition table saved in module B or Weights entered in module C will not be deleted.

1 Because the computer represents all numbers as real numbers it is possible that a number with 5 as the third decimal and 0 as the fourth and subsequent decimals is displayed on the screen in some cases after rounding up and in some cases after rounding down. Since all calculations are performed with a minimum of four decimals this does not influence the results.

PROCEDURE A1.1. DISCOUNT FUTURE EFFECTS

Purpose of this procedure

This procedure can be used to transform a series of effects tables that include effect scores for future years to a single discounted effects table that includes discounted values for a common base year.

Your input and DEFINITE's output

Enter:

1. The **base year**: this is usually the year of the first expenditure on a project;
2. **Effect scores** for the base year;
3. The **next year**;
4. **Effect scores** for the next year;
5. Repeat this for all following years;
6. One or more **discount rates.**

This results in:

A discounted effects table.

Example

Your input

Enter:

1. The **base year**: In this example **1988;**
2. The **effect scores** for the base year;

Effects table (base year 1988). Improve transportation to Nyverhoeck.

effects	units	Highway	Road/bus	Train
Costs	millions	-200	-250	-400
Travel time	min/prs/day	30	15	10
Capacity	mln/km/yr	20	30	40
Emissions NO_x	ton/yr	-1000	-750	-100
Landscape	hectare	-10	-12	-2

Note that Travel time reflects the reduction in travel time (benefit) and Landscape reflects the loss of landscape (cost) that result from the alternatives.

3. The **next year**: **1989**;
4. The **effect scores** for 1989;

Effects table (1989). Improve transportation to Nyverhoeck.

effects	units	Highway	Road/bus	Train
Costs	millions	-20	-25	-40
Travel time	min/prs/day	30	15	10
Capacity	mln/km/yr	20	30	40
Emissions NO_x	ton/yr	-1000	-750	-100
Landscape	hectare			

5. There are no more following years;
6. A uniform **discount rate** of 10%.

DEFINITE's output

Discounted effects table. Improve transportation to Nyverhoeck.

effects	units	Highway	Road/bus	Train
Costs	millions	-218.18	-272.73	-436.36
Travel time	min/prs/day	57.27	28.64	19.09
Capacity	mln/km/yr	38.18	57.27	76.36
Emissions NO_x	ton/yr	-1909.09	-1431.82	-190.91
Landscape	hectare	-10.00	-12.00	-2.00

In this example all effects are discounted at the same rate. It is also possible to use a different rate for each effect.

Comments

Calculation of the discounted values

The standard effects table is the effects table in the base year (year 0). Costs and benefits that occur after the base year are included in time-effects tables (time.***). These time-effects tables are linked to specific years. A maximum of 50 time-effects tables can be entered.

Costs and benefits that occur during a certain year are attributed to the end of that year. Figure 4.2 shows that costs and benefits that occur in year 2 are attributed to the end of year 2 before being discounted over 2 years to the base year. Costs and benefits occurring in the base year are not discounted. This implies that the investment decision is assumed to take place on the first of January of year 1. It is assumed implicitly that costs and benefits in year 0 occur at the end of this year. Costs in year 0 are usually investment costs and there are usually no benefits. If the investment costs are spread over a longer period it may be preferable also to discount the investment costs. The year of investment should then be entered as year 1 and the base year as the year before. In this case the effects table of the base year should be empty.

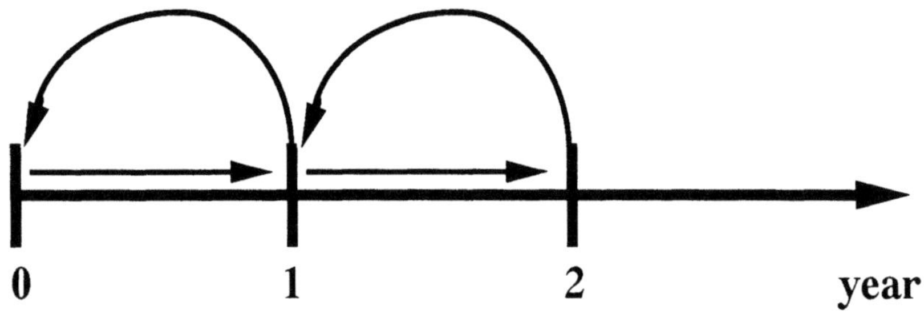

Figure 4.2. Effects tables, time and discounting.

Figure 4.3 shows this procedure for the example that was shown above. Year 0 is now replaced by 1988 (the base year) and year 1 now equals 1989.

The equation used to discount future costs and benefits is shown below.

$$x_{ji}^{d} = \sum_{t^0}^{t^n} x_{ji}^{t} * \frac{(1+p_i/100)^{t-t^0}}{(1+r_i/100)^{t-t^0}}$$

In this equation x_{ji}^{d} represents the discounted effect i of alternative j and x_{ji}^{t} represents this effect at year t. The base year is shown as t^0 and the last year to be included (the planning horizon) as t^n. In this equation r_i represents the discount rate and as p_i the price index, both entered as percentages. The price index can be used to correct for changes in prices over time. The discount rate can be specified as a uniform discount rate for all effects or can be specified for each effect separately.

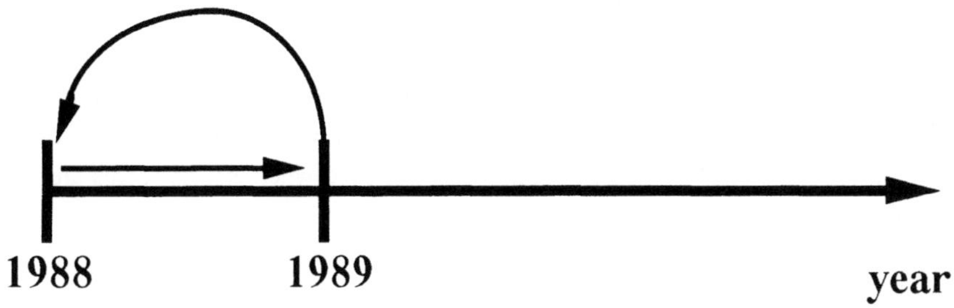

Figure 4.3. Effects tables, time and discounting.

Qualitative scores

Qualitative scores (ordinal and ---/+++) cannot be discounted and can therefore only be entered in the base year.

PROCEDURE A1.2. ASSESS SCORES

Purpose of this procedure

This procedure generates indices that represent the relative magnitudes of one effect in relation to all alternatives. The values of the indices for one effect result from comparison of all pairs of alternatives for this effect. This procedure can be used when effects scores are not (yet) available.

Your input and DEFINITE's output

Enter:

1. The **effect** for which the scores need to be assessed;
2. A **comparison** of the relative qualities of each pair of alternatives for this effect.

This results in:

A row of index values for this effect.

Example

This example begins with an empty effects table.

Effects table. Improve transportation to Nyverhoeck.

effects	units	Highway	Road/bus	Train
Costs	mln $			
Travel time	min/prs/day			
Capacity	mln/km/yr			
Emissions NO_x	ton/yr			
Landscape	hectare			

Your input

Enter:

A **comparison** of each pair of alternatives. In this example the alternatives are compared on their effects on landscape.

Highway is slightly better than Road/bus
Train is much better than Highway
Train is very much better than Road/bus

The other effects have to be assessed in a similar way.

DEFINITE's output

Effects table. Improve transportation to Nyverhoeck.

effects	units	Highway	Road/bus	Train
Costs	index			
Travel time	index			
Capacity	index			
Emissions NO_x	index			
Landscape	index	0.26	0.11	1.00

inconsistency index = 0.054

Comments

An inconsistency ratio is presented.[1] This ratio indicates whether the pairwise comparisons are sufficiently consistent. A value of 0 indicates complete consistency. An inconsistency ratio of 0.1 or more indicates inconsistent judgements.

---

1 See: Saaty T.L., and L.G. Vargas (1984). Comparison of Eigenvalue and logarithmic least squares and least squares methods in estimating ratios. *Mathematical modelling*, Vol. 5, pp. 309-324.

PROCEDURE A2. GENERATE A LIST OF EFFECTS

Purpose of this procedure

This procedure offers a checklist of effects. This checklist is divided into 11 categories. The purpose of this procedure is to use this checklist to generate a list of effects that is as complete as possible.

Enter:

1. The main **categories** that are relevant to your problem, e.g. Environment and Employment;
2. The **effects from the checklist** to be included in the effects table;
3. **Other effects** to be included in the effects table;
4. The **measurement scale, measurement unit** and **cost** or **benefit** characteristic of the included effects.

This results in:

A list of effects.

By selecting **continue** you enter procedure A1: **create an effects table** to complete the effects table.

Example

Your input

Enter:

1. Main **categories**. In this example the main categories of effects which are considered of importance are: costs, environment and mobility;
2. DEFINITE presents a checklist of **effects** for every main category. The user selects distance to work, distance to public transport and traffic-jam from the main category: mobility;
3. The user adds the **other effects** maximum speed and transport costs;
4. **Measurement level, measurement unit, cost** or **benefit.**

DEFINITE's output

A list of effects.

effects	measurement units	Cost or Benefit
Distance to work	km/prs	Cost
Distance to public transport	km/prs	Cost
Traffic-jam	ordinal	-
Maximum speed	km/hr	Benefit
Transport costs	x 1000 $	Cost

This list shows only effects for the mobility category.

PROCEDURE A3. GENERATE ALTERNATIVES

Purpose of this procedure

The purpose of this procedure is to generate a list of alternatives that is as complete as possible. This procedure can only be used if it is possible to:

- distinguish various aspects or dimensions of the problem, and
- per aspect one or more solutions can be identified.

Your input and DEFINITE's output

Enter:

1. Two or more **aspects** of the problem;
2. One or more **solutions** for each aspect;
3. **Feasible and not feasible combinations.**

This results in:

A list of all feasible combinations of aspects and solutions.

By selecting **continue** you enter procedure A1: **create an effects table** to complete the effects table (see the map of the region included in Chapter 2).

Example

In this example the problem is to provide a connection across the river Traach (see the map of the region included in Chapter 2).

Your input

Enter:

1. **Aspects** of the problem, in this example the means of transport, the type of connection and its capacity;
2. **Solutions** per aspect, for example train or car for the aspect means of transport;

Aspects and solutions. Crossing the river Traach.

Aspects	Solutions		
Means of transport	Train	Car	
Type of connection	Ferry	Tunnel	Bridge
Capacity (persons/day)	10000	100000	

3. **Feasible and non-feasible combinations.** The combinations train - ferry and ferry - 100,000 persons/day are considered non-feasible.

DEFINITE's output

A list of alternatives: each alternative consists of some combination of the solutions shown above. Because train - ferry and ferry - 100,000 persons/day are considered non-feasible the alternatives beginning with AA. or ending with .AB are not present in this table. A selection of the generated alternatives is shown below:

Alternatives. Crossing the river Traach.

Alt ABA	Alt ABC	Alt ACA	....	Alt BCB
Train	Train	Train		Car
Tunnel	Tunnel	Bridge		Bridge
10000	100000	10000		100000

Comments

The number of alternatives generated increases combinatorially with the number of solutions per specified aspect. If a large number of solutions is available the specification of non-feasible combinations becomes more important. At most 50 alternatives can be stored for further use when you leave this procedure.

From this procedure you can continue to procedure A2 to generate a list of effects or you can continue directly to procedure A1 to enter effects and effect scores.

PROCEDURE A4. COMBINE ALTERNATIVES

Purpose of this procedure

The purpose of this procedure is to combine elements or parts of alternatives into alternatives. This procedure is only relevant if:

- elements exist that can be combined into alternatives;
- elements are not mutually exclusive;
- effects of elements can be aggregated.

Your input and DEFINITE's output

Enter:

1. Names of the **elements**;
2. Names of the **descriptors** and **effects**. For each effect or descriptor you need to enter: the name of the effect or descriptor, the measurement scale (in this procedure only a ratio scale is permitted), the measurement unit e.g. m^3/yr and costs or benefit;
3. **Scores** for the effects and descriptors;
4. **Combination rules.** These are minimum and maximum values that the alternatives must meet e.g. a minimum capacity or a maximum budget;
   Combination rules can be defined on effects and descriptors;
   Combination rules can also be defined on nominal descriptors e.g. a maximum of 2 elements in region A and a minimum of 3 elements in region B.

This results in:

1. An effects table;
2. A table showing the composition of the alternatives.

No more than 50 alternatives can be generated.

Example

To secure the water supply of Deil aan Zee a minimum of 10,000 m^3/yr of water needs to be extracted from ground water stocks. Four wells are available in the dune area nearby. These wells, the elements of an alternative, have capacities of 3000, 5000, 2000 and 6000 m^3/yr (see the map of the region included in Chapter 2).

Your input

Enter:

1. **Elements:** Well1, Well2, Well3, Well4;
2. **Descriptors** and **effects**: Capacity (m^3/yr), Land use (m^2, cost), Ground water level (cm, cost);
3. **Scores;**

This results in an elements table as shown below. Note that both effects are costs and therefore recieve a negative sign.

Elements table. Water supply of Deil aan Zee.

		Well1	Well2	Well3	Well4
Capacity	m^3/yr	3000	5000	2000	6000
Land use	m^2	-200	-400	-200	-700
Ground water level	cm	-5	-10	-4	-15

4. **Combination rule**. In this example each alternative must produce at least 10,000 m^3/year and no more than 12,000 m^3/year.

DEFINITE's output

Effects table. Water supply of Deil aan Zee.

		Alt.1	Alt.2	Alt.3
Capacity	m^3/yr	10000	11000	11000
Land use	m^2	-800	-1100	-1100
Ground water level	cm	-19	-24	-25

This procedure also shows the composition of the alternatives.

In this example:

Alternative 1 = Well1 + Well2 + Well3

Alternative 2 = Well1 + Well3 + Well4

Alternative 3 = Well2 + Well4

Comments

The number of alternatives generated increases combinatorially with the number of elements. If a large number of elements is available the specification of combination rules becomes more important. At most 50 alternatives can be stored for further use when you leave this procedure.

PROCEDURE A5. DISCARD IRRELEVANT ALTERNATIVES

Purpose of this procedure

This procedure selects alternatives that are not relevant to the decision problem. An alternative can be not relevant if it does not meet certain hard constraints such as available budget or environmental standards. An alternative may also be irrelevant if at least one other alternative can be found that is better than this alternative for at least one effect and as good as or better for all other effects. Such an alternative is usually referred to as an inefficient or dominated alternative.

This procedure is especially useful for problems with a large number of alternatives. Such a situation can result from procedure A4. This procedure is not useful if a full ranking of the alternatives is required.

Constraints

Your input and DEFINITE's output

Enter:

Constraints for one or more effects.

This results in:

Highlighting of alternatives that do not meet the constraints.

Inefficient alternatives

Your input and DEFINITE's output

Enter:

<F1> inefficient

This results in:

1. Highlighting of inefficient alternatives;
2. Identification of alternatives that dominate the inefficient alternatives.

Example

Effects table. Improve transportation to Nyverhoeck.

effects	units	Highway	Road/bus	Train	Air
Costs	mln $	-200	-250	-400	-800
Travel time	min/prs/day	30	15	10	10
Capacity	mln/km/yr	20	30	40	10
Emissions NO_x	ton/yr	-1000	-750	-100	-2000
Landscape	hectare	-10	-12	-2	-14

Note that Travel time reflects the reduction in travel time (benefit) and Landscape reflects the loss of landscape (cost) that result from the alternatives.

Constraints

Your input

Enter:

Constraints. In this example the user indicates an upper limit to costs of 300 mln $, and to Emissions NO_x of 900 ton/year.

DEFINITE's output

DEFINITE selects the alternatives which do not meet these constraints.

Irrelevant alternatives. Improve transportation to Nyverhoeck.

effects	units	**Highway**	Road/bus	**Train**	**Air**
Costs	mln $	-200	-250	-400	-800
Travel time	min/prs/day	30	15	10	20
Capacity	mln/km/yr	20	30	40	10
Emissions NO_x	ton/yr	-1000	-750	-100	-2000
Landscape	hectare	-10	-12	-2	-14

There is only one alternative which fulfils both conditions. If the question: 'Do you want to remove the marked alternatives from the effects table?' is answered with yes, the resulting effects table will only include one alternative.

Inefficient alternatives

Your input

With the option: **inefficient**, the inefficient alternatives will be selected and the effects table becomes:

DEFINITE's output

Dominated alternatives. Improve transportation to Nyverhoeck.

effects	units	Highway	Road/bus	Train	Air
dominated by					Highway
Costs	mln $	-200	-250	-400	-800
Travel time	min/prs/day	30	15	10	10
Capacity	mln/km/yr	20	30	40	10
Emissions NO_x	ton/yr	-1000	-750	-100	-2000
Landscape	hectare	-10	-12	-2	-14

This shows Air as an inefficient alternative because it is in all respects worse than, and is therefore dominated by, Highway.

5. PROBLEM PRESENTATION PROCEDURES

Introduction

The procedures available in **module B: problem presentation** can be used to present the effects table in a form that supports comparison of the alternatives without applying a formal decision rule. Problem presentation is supported by the following procedures:

B1. **Method information** to support the selection of an evaluation method that matches the characteristics of an effects table.

B2/B3. **Define** and **calculate appraisal table** to transform an effects table into a smaller appraisal table. [This method transforms an effects table into a smaller appraisal table.] An appraisal table includes a maximum of 10 appraisal categories. You need to specify coefficients that link effects to categories. Using these coefficients the appraisal scores are calculated as the weighted sum of the (standardized) effects scores.

B4. **Presentation and ranking** to present and rank an effects or appraisal table graphically.

B5. **Cost-benefit sheet** to assign prices to effects and to transform an effects table into a cost-benefit sheet. This is a necessary step to perform Cost-Benefit analysis in the next module.

B6. **Presentation** to provide a graphic presentation of the benefit-cost ratios.

B7. **Diagram** to provide a scatter diagram and ranking of two categories selected from the appraisal table.

A flow diagram of the procedures of **module B: problem presentation** is shown in Figure 5.1.

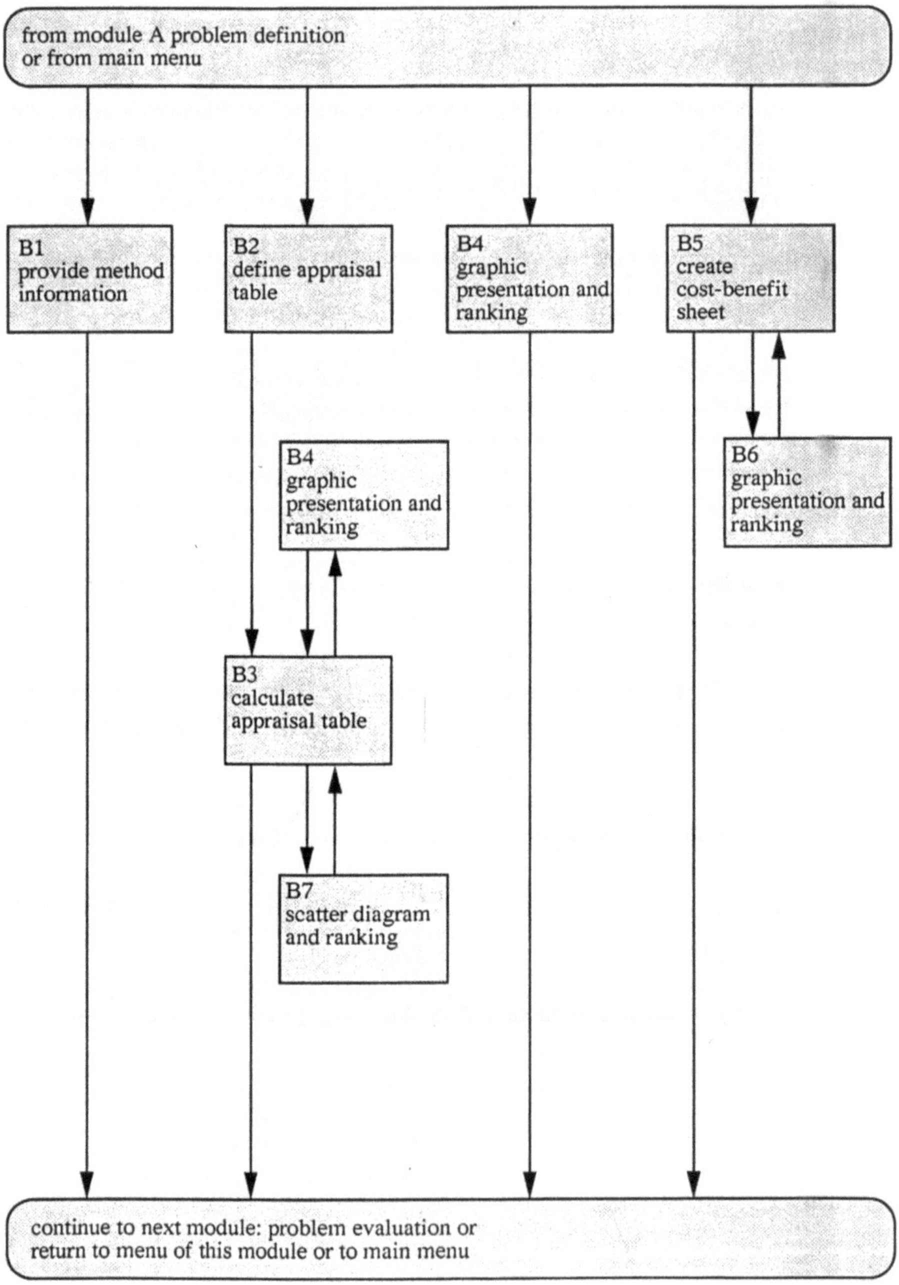

Figure 5.1. Procedures module B: problem presentation.

DEFINITE menus

Upon entering the module, you will see the module menu. This menu provides access to the procedures available in this module. Each procedure is executed through its command line at the bottom of the screen.

There are two ways to make a selection from a menu:

1. Press **<SPACE>** to move through the menu and hit **<RETURN>** to select the highlighted option. An information line above the command line provides information on each highlighted option;
2. Hit the function key that corresponds to the option you want to select (e.g. **<F1>** to select the first option in a menu).

In every command line you can:

- Select **continue** to complete a procedure and to continue to the next step in the program;
- Hit **<ESC>** to cancel a procedure and return to the previous step;
- The **<CURSOR>** or **<ARROW>** keys are used to move the cursor within a table.

PROCEDURE B2/B3. APPRAISAL TABLE

Purpose of this procedure

This procedure transforms an effects table into a smaller appraisal table. An appraisal table includes a maximum of 10 appraisal categories. You need to specify coefficients that link effects to categories. Using these coefficients the appraisal scores are calculated as the weighted sum of the effects scores. This is the default setting.

To combine scores measured in different units or even different measurement scales these scores need to be standardized before they can be aggregated. If this is required you need to go to Utilities from the main menu to change the default setting to the setting you want.

Your input and DEFINITE's output

Enter:

1. The **number** of categories;
2. **Names** and **measurement units** of these categories;
3. **Coefficients** to link effects to categories.

This results in:

1. A definition table of coefficients;
2. An appraisal table.

If an appraisal table is available the multicriteria methods in problem evaluation automatically select this table.

Example

Effects table. Improving a highway.

effects	units	Alt.1	Alt.2	Alt.3
Maintenance Costs	x 1000 $	-40	-60	-80
Toll collection	x 1000 $	0	80	100
Reduced travel time	minutes	25	0	20
Loss of landscape	ordinal	3	1	2
Decrease fatal accidents	number/yr	5	7	8
Decrease accidents with wounded	number/yr	10	15	20
Decrease accidents with damage	number/yr	50	100	150

Your input

Enter:

1. The **number** of categories: e.g. 3;
2. **Names** and **measurement** units of these categories. In this example: profits, accidents and landscape damage;
3. **Coefficients** to link effects to categories. For example: fatal accidents are valued 5 times higher than accidents that result in people being wounded but not killed and valued 10 times higher than accidents that result in material damage only.

DEFINITE's output

Definition table. Improving a highway.

Profits	=	Toll collection - Maintenance costs
Decrease accidents	=	10x Fatalacc. + 5x Acc. with wounded- + Acc. with damage
Landscape damage	=	Loss of landscape

Appraisal table. Improving a highway.

category	units	Alt.1	Alt.2	Alt.3
Profits	x 1000 $	-40	20	20
Decrease accidents	index	150	245	330
Landscape damage	ordinal	3	1	2

Comments

Clean-up

The appraisal table calculated in this procedure is input to the multicriteria methods available in module C. Multicriteria analysis cannot be applied in one session on both the effects table and the appraisal table. If you want to apply multicriteria analysis within this session on the effects table you will have to delete the appraisal table using the clean-up procedure when you leave this module.

Positive and negative values

A category can have a positive or negative score depending on the values of the coefficients and the effects. Therefore, in contrast to the effects table, in an appraisal table positive and negative values can occur in one row.

PROCEDURE B4. PRESENTATION AND RANKING

Purpose of this procedure

This procedure presents the effects table or the appraisal table graphically. Scores are standardized between zero and one. For both costs and benefits a higher column indicates a better score.

You can order the presentation by exchanging rows and columns and by selecting the **rank** option.

Your input and DEFINITE's output

Enter:

1. Importance of effects by exchanging rows;
2. **Rank** to order the alternatives.

This results in:

A first priority ordering of the alternatives.

Example

Effects table. Improve transportation to Nyverhoeck.

effects	units	Highway	Road/bus	Train
Costs	mln $	-200	-250	-400
Travel time	---/+++	+++	++	+
Capacity	mln/km/yr	20	30	40
Emissions NO_x	ton/yr	-1000	-750	-100
Landscape	---/+++	---	---	-

Your input

Enter:

This procedure requires no new input.

DEFINITE's output

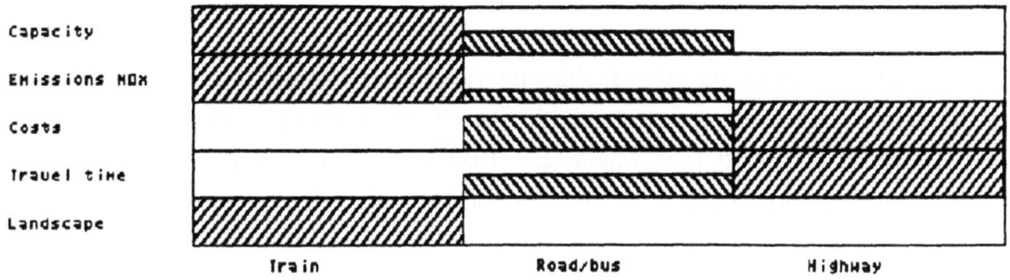

Figure 5.2. Graphic presentation of the ranked effects table.

In this presentation the higher the column the better the alternative. Therefore an empty block marks the worst alternative and a full block marks the best alternative. In ranking the alternatives the effect in the top row is interpreted as the most important effect of the problem, followed by the effect in the second row, and so on.

Comments

Ranking the alternatives

If you select rank DEFINITE transforms the order of rows to quantitative weights using the Expected Value method (see procedure C3) and then ranks the columns using Weighted Summation (see procedure C2.1).

Printing the result

Figures are *not* included in the report to be generated in module E. It is therefore advisable to print this figure now. Procedures to save screen output are available for including figures in a text file. Users of WordPerfect 5.0 or higher, for example, can use the GRAB program to include this figure in a WordPerfect document. See the WordPerfect manual for further instructions.

PROCEDURE B5. COST-BENEFIT SHEET

Purpose of this procedure

This procedure transforms the effects table into a cost-benefit sheet for each alternative. You need to supply a price for as many quantitative effects as possible. The effect scores are multiplied by these prices to obtain monetary effect scores. Effects that cannot be priced are included as *pro memoria* (p.m.) in the cost-benefit sheet. If you have not used the **time** option in module A (to specify effects tables for more than one year followed by discounting) you can do this, if this is relevant to your problem, on entering this procedure. Only two alternatives can be displayed simultaneously. Use print to display all cost-benefit sheets.

Your input and DEFINITE's output

Enter:

1. **Prices**

This results in:

1. A cost-benefit sheet for each alternative;
2. A graphic presentation of the benefit-cost ratios.

A cost-benefit sheet is a necessary input to Cost-Benefit analysis or to Cost-Effectiveness analysis in the next module problem evaluation.

Example

The discounted effects table below serves as an example (see also procedure A1.1).

Discounted effects table. Improve transportation to Nyverhoeck.

effects	units	Highway	Road/bus	Train
Costs	millions	-218.18	-272.73	-436.36
Travel time	min/prs/day	57.27	28.64	19.09
Capacity	mln/km/yr	38.18	57.27	76.36
Emissions NO_x	ton/yr	-1909.09	-1431.82	-190.91
Landscape	hectare	-10.00	-12.00	-2.00

Note that Travel time reflects the reduction in travel time (benefit) and Landscape reflects the loss of landscape (cost) that result from the alternatives.

This procedure converts the above effects table into the cost-benefit table below. In this table the effects are specified as costs or benefits. Monetary scores, in this case costs only, do not need a price.

Prices which have not yet been assigned are indicated by a '?'. Notice the fact that all prices are expressed in multiples of millions.

Cost-benefit table. Improve transportation to Nyverhoeck.

effects	Cost Benefit	units	prices millions	Highway	Road/bus	Train
Costs	C	millions	1	218.18	272.73	436.36
Emissions NO_x	C	ton/yr	?	1909.09	1431.82	190.91
Landscape	C	hectare	?	10.00	12.00	2.00
Travel time	B	min/prs/day	?	57.27	28.64	19.09
Capacity	B	mln/km/yr	?	38.18	57.27	76.36

Your input

Enter:

Prices. In this example loss of landscape is assigned a price of 10 million per hectare and reduced travel time is assigned a price of 4 million per minute. Emissions NO_x and capacity are not priced.

DEFINITE's output

After selecting: **C/B sheet** the cost-benefit table is first transformed into the table below.

Monetary cost-benefit table (scores in millions). Improve transportation to Nyverhoeck.

effects	Cost Benefit	units	prices millions	Highway	Road/bus	Train
Costs	C	millions	1	218.18	272.73	436.36
Emissions NO_x	C	ton/yr	?	p.m.	p.m.	p.m.
Landscape	C	hectare	10	100.00	120.00	20.00
Travel time	B	min/prs/day	4	229.08	114.56	76.36
Capacity	B	mln/km/yr	?	p.m.	p.m.	p.m.

Note that Travel time reflects the reduction in travel time (benefit) and Landscape reflects the loss of landscape (cost) that result from the alternatives.

After selecting **continue** this table is converted to cost-benefit sheets.

Cost-benefit sheets. Improve transportation to Nyverhoeck.

Highway millions				Road/bus millions			
Costs		Benefits		Costs		Benefits	
Costs	218.18	Travel time	229.08	Costs	272.73	Travel time	114.56
Landscape	100.00	Capacity	p.m.	Landscape	120.00	Capacity	p.m.
Emissions NO_x	p.m.			Emissions NO_x	p.m.		

Comments

Some CBA conventions require that costs that are negative effects of alternatives (unintended costs), such as the loss of landscape, are listed as negative benefits instead of costs. Use the option **negative benefits** to transform costs into negative benefits.

PROCEDURE B6. GRAPHIC PRESENTATION OF BENEFIT-COST RATIOS

Purpose of this procedure

This procedure provides a graphic presentation of the benefit-cost ratios.

Example

The monetary cost-benefit table as calculated in procedure B5 serves as an example.

Monetary cost-benefit table (scores in millions).

effects	Cost Benefit	units	prices millions	Highway	Road/bus	Train
Costs	C	millions	1	218.18	272.73	436.36
Travel time	B	min/prs/day	4	229.08	114.56	76.36
Capacity	B	mln/km/yr	?	p.m.	p.m.	p.m.
Emissions NO_x	C	ton/yr	?	p.m.	p.m.	p.m.
Landscape	C	hectare	10	100.00	120.00	20.00

Note that Travel time reflects the reduction in travel time (benefit) and Landscape reflects the loss of landscape (cost) that result from the alternatives.

DEFINITE's output

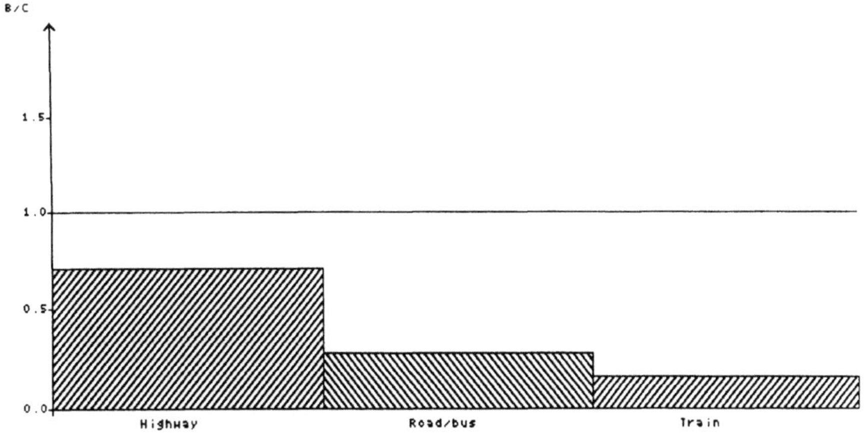

Figure 5.3. Benefit-cost ratios.

Comments

Printing the result

Figures are *not* included in the report that can be generated in module E. It is therefore advisable to print this figure now. Procedures to save screen output are available for including figures in a text file. Users of WordPerfect 5.0 or higher, for example, can use the GRAB program to include this figure in a WordPerfect document. See the WordPerfect manual for further instructions.

PROCEDURE B7. SCATTER DIAGRAM

Purpose of this procedure

This procedure generates a scatter diagram of the scores of two categories selected from the appraisal table. Scores are standardized between 0 and 100. For both costs and benefit scores, a score of 100 represents the best score. This procedure is useful to compare many alternatives on a few categories.

You may create a ranking of the alternative based on the two selected categories only. The alternative which is closest to the upper right corner ranks first, the second closest ranks second, etc. The ranking can be derived graphically by selecting **move line**. **Change Wx** changes the weight of the category on the horizontal axis and therefore also the weight on the vertical axis since these weights add up to one.

Your input and DEFINITE's output

Enter:

1. The **categories** to be compared;
2. **Change Wx** to change the relative weights of the categories;
3. **Move line** to move a line through the diagram.

This results in:

1. A diagram that includes scores for two categories on all alternatives;
2. A ranking of the alternatives based on these two categories;
3. The correlation between these categories. A value close to one indicates minimum conflict between the categories and therefore minimal sensitivity to the weights. A value close to minus one indicates the opposite.

Example

The appraisal table calculated in procedure B3 serves as an example.

Appraisal table. Improve transportation to Nyverhoeck.

categories	units	Alt.1	Alt.2	Alt.3
Profits	x 1000	-40	20	20
Decrease accidents	index	150	245	330
Landscape	ordinal	3	1	2

Your input

Enter:

1. **Categories** to be compared. In this example: profits and decrease in accidents;
2. **Weights.** In this example: profits 0.5 and therefore accidents also 0.5;
3. **Move line.**

DEFINITE's output

A diagram and a ranking based on the scores of two categories.

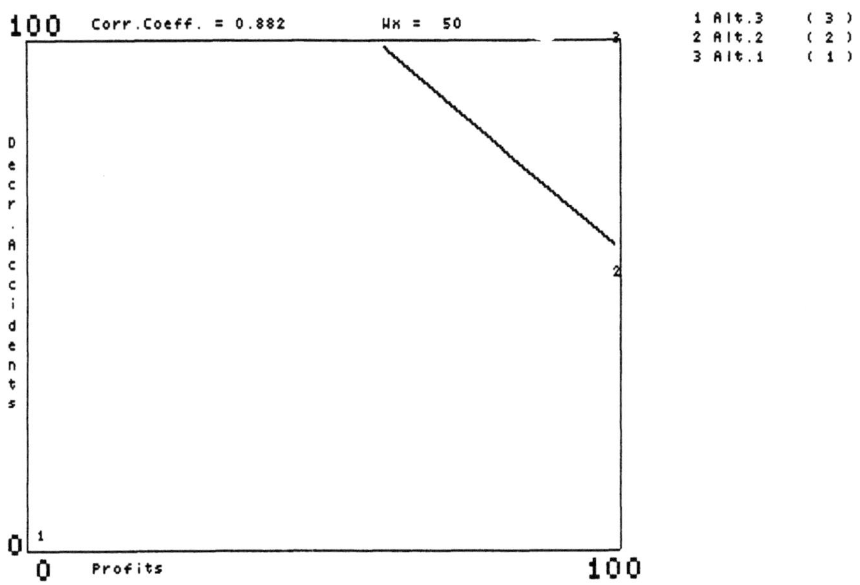

Figure 5.4. Scatter diagram comparing profits against the decrease in accidents.

Comments

This procedure is only useful for comparing many alternatives on a few appraisal categories. Procedure B5: **graphic presentation** is more suited to problems with few alternatives and many criteria. The correlation coefficient provides an indication of the sensitivity of the ranking to changes in weights; a value of 1 indicates complete insensitivity and a value of -1 maximum sensitivity.

Printing the result

Figures are *not* included in the report that can be generated in module E. It is therefore advisable to print this figure now. Procedures to save screen output are available for including figures in a text file. Users of WordPerfect 5.0 or higher, for example, can use the GRAB program to include this figure in a WordPerfect document. See the WordPerfect manual for further instructions.

6. PROBLEM EVALUATION PROCEDURES

Introduction

The procedures available in **module C: problem evaluation** can be used to rank alternatives. DEFINITE contains two main categories of methods: multicriteria methods and monetary methods.

Problem evaluation is supported by the following procedures.

C1. **Method information** to support the selection of an evaluation method that matches the characteristics of an effects table.

C2.1. **Weighted Summation** is a simple and frequently used multicriteria method. This method generates a ranking of the alternatives. As a first step all effects are standardized. An appraisal score is then calculated for each alternative by first multiplying each standardized effect score by its appropriate weight followed by summing the weighted scores of all effects.

C2.2. **Electre 2** calculates the ranking of alternatives based on pairwise comparison of alternatives. Both the degree to which one alternative is better than other alternatives (concordance) and the degree to which it is worse than other alternatives (discordance) are used. This information is used to generate a ranking of the alternatives. This ranking is sometimes incomplete.

C2.3. The **Regime method** uses a random generator for each pair of alternatives to estimate the probability that one alternative ranks above the other. Large effects tables may lead to some waiting time on slow computers.

C2.4. The **Expected Value method** generates a ranking of the alternatives. As a first step all effect scores are standardized according to the Expected Value method. An appraisal score is then calculated for each alternative by first multiplying each standardized effect score by its appropriate weight followed by summing the weighted scores of all effects.

C2.5. The **Evamix method** generates a ranking of the alternatives. The Evamix method generates separate dominance tables for quantitative and for qualitative effect scores. Both tables are combined into a total dominance table.

C3. **Create weights.** Multicriteria analysis requires information on priorities (weights). This procedure includes various procedures to transform quantitative or qualitative priority statements into quantitative or qualitative weights.

C4. **Presènt results** presents the results of multicriteria analysis graphically.

C5. **Standardization** If the default standardization is considered inappropriate this procedure can be used to select a specific method of standardization for the effects table as a whole, or for each effect individually.

C6.1. **Cost-Benefit analysis** calculates the net present value, the benefit-cost ratio and the internal rate of return for each of the alternatives. This procedure uses the cost-benefit table created in module B. This procedure is useful in problems where most effects can be valued in monetary terms.

C6.2. **Cost-Effectiveness analysis** can be used in three ways:

1. To determine which alternative, given a certain level of effects, can be realized as efficiently as possible (cost minimization);
2. To determine which alternative, given a certain level of costs, can be realized as efficiently as possible (effect maximization);
3. To determine which alternative should be realized, given its effect divided by its costs (cost effectiveness).

A flow diagram of the procedures of **module C: problem evaluation** is shown in Figure 6.1.

DEFINITE menus

Upon entering the module, you will see the module menu. This menu provides access to the procedures available in this module. Each procedure is executed through its command line at the bottom of the screen.

There are two ways to make a selection from a menu:

1. Press the **<SPACE>** bar to move through the menu and hit **<RETURN>** to select the highlighted option. An information line above the command line provides information on each highlighted option.
2. Hit the function key that corresponds to the option you want to select (e.g. **<F1>** to select the first option in a menu).

In every command line you can:

- Select **continue** to complete a procedure and to continue to the next step in the program;
- Hit **<ESC>** to cancel a procedure and return to the previous step;
- The **<CURSOR>** or **<ARROW>** keys are used to move the cursor within a table.

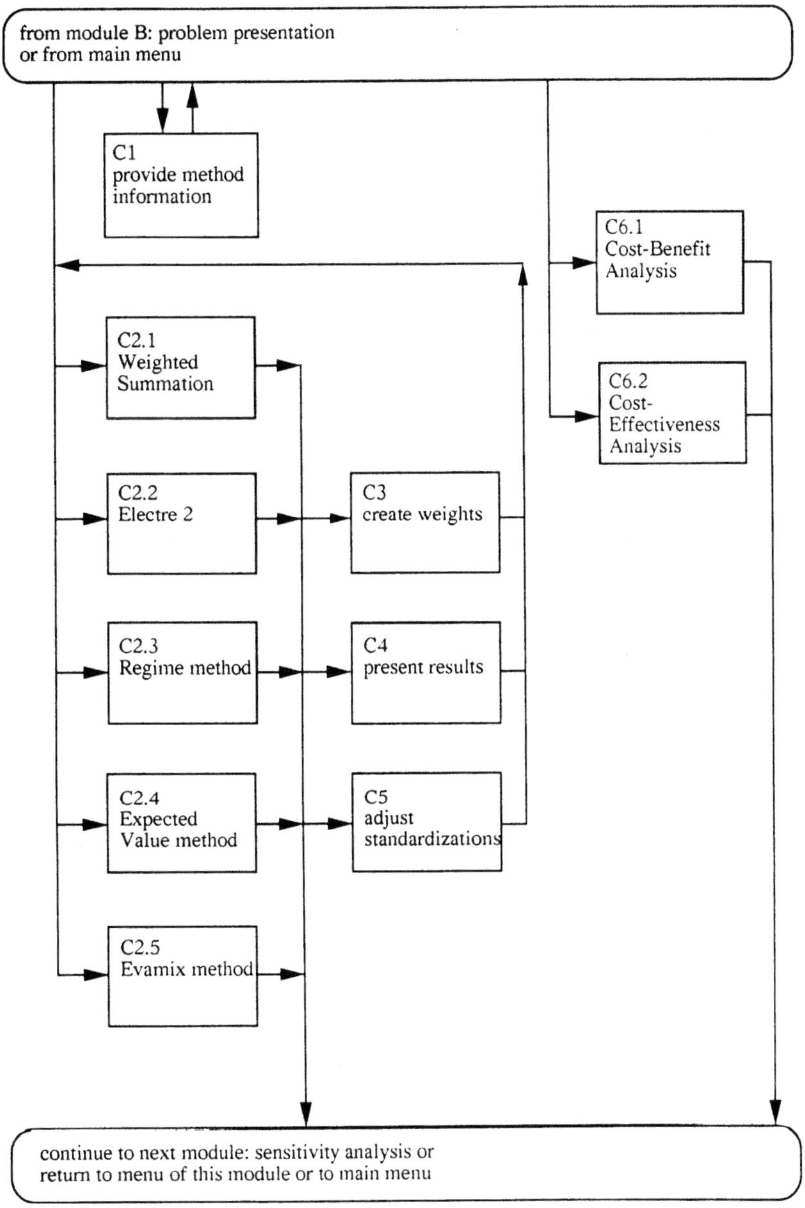

Figure 6.1. Procedures module C: problem evaluation.

PROCEDURE C2.1. WEIGHTED SUMMATION

Purpose of this procedure

This procedure generates a ranking of the alternatives.

Description

Weighted Summation is a simple and frequently used evaluation method. As a first step all effect scores are standardized. An appraisal score is then calculated for each alternative by first multiplying these standardized effect scores by its appropriate weight followed by summing the weighted scores of all effects.

Your input and DEFINITE's output

Enter:

1. **Weights** (if not yet available);
2. **Standardization** (if changes to the default standardization are required);
3. **<F1>** to execute the method.

This results in:

1. A ranking of the alternatives;
2. An appraisal score for each alternative.

This procedure calculates the ranking of alternatives based on the weighted sum of the standardized effect scores. If you have not yet assigned weights, you need to do so within this procedure.

Example

Effects table. Improve transportation to Nyverhoeck.

effects	units	Highway	Road/bus	Train
Costs	mln $	-200	-250	-400
Travel time	min/prs/day	30	15	10
Capacity	mln/km/yr	20	30	40
Emissions NO_x	ton/yr	-1000	-750	-100
Landscape	hectare	-10	-12	-2

Your input

Enter:

1. **Weights** (see also procedure C3):

Costs	0.2
Travel time	0.1
Capacity	0.3
Emissions NO_x	0.3
Landscape	0.1

2. Do not select standardization in order to use default standardization;
3. **Execute.**

DEFINITE's output

The table below, which shows the weights, the ranking, and the scores.

Results of Weighted Summation. Improve transportation to Nyverhoeck.

weights rank		score	ranking rank	score
1.	Capacity	0.300	1. Train	0.70
	Emissions NO_x	0.300	2. Road/bus	0.41
3.	Costs	0.200	3. Highway	0.32
4.	Travel time	0.100		
	Landscape	0.100		

Comments

Printing the result

At most 16 alternatives and/or effects are shown on the screen. If your evaluation problem includes more alternatives and/or effects you need to print the table to see the complete result.

Saving the result

It is recommended to save each relevant result of a multicriteria method. A result that is saved can be analyzed using sensitivity analysis in module D and will be included in the report generated in module E. When you save a result you are asked whether you want to overwrite the previous result. This question is asked to prevent saving irrelevant results. Answer **no** if you want to save both results.

To be able to identify the various results to be analyzed in module D each result receives a name and information line. For example, **summ.001** identifies the first result of Weighted Summation that was saved.

The information line can be used to further identify the result. As default the information line provides the weight method used and the value of the highest weight. This information line can be replaced or extended. The information line: 'Direct: capacity = 0.400, reference set' indicates that this result was obtained using direct assessment of weights, capacity has weight 4.00 and the weight set was labelled 'reference set'.

Standardization

Scores and ranking are dependent on the standardization method applied. The results of different calculations can be compared by saving the MCA results. In this way the influence of changes in weights, weight methods and standardization procedures can be analyzed.

If procedure C5, **standardization**, is *not* selected each row of effect scores is automatically standardized as follows:

- monetary, ratio,
  interval and
  ordinal scale : (score-lowest score) / (highest score-lowest score);
- binary scale : yes = 1; no = 0 *or* yes = 0; no = 1;
- ---/+++ scale : ---, --, -, 0, +, ++, +++ = 0, 1/6, 2/6, 3/6, 4/6, 5/6, 1.

PROCEDURE C2.2. ELECTRE 2 METHOD

Purpose of this procedure

This procedure calculates the ranking of alternatives based on pairwise comparison of alternatives. If you have not yet assigned weights, you need to do so within this procedure.

Description

This method, also known as Concordance analysis, is widely used and is especially popular in French-speaking countries. The version described below is known as Electre 2. The Electre method is based on a pairwise comparison of the alternatives, thus using only the interval character of the scores in the evaluation of the effects table. The basic idea is to measure the degree to which scores and their associated weights confirm or contradict the dominant pairwise relationships among alternatives.

A dominance relationship for each pair of alternatives is derived using both an index of concordance and an index of discordance. The concordance index represents the degree to which alternative i is better than alternative i'. This index is defined as the sum of weights of the effects included in the concordance set $C_{ii'}$. This is the set of effects for which alternative i is at least equally attractive as alternative i'.

The discordance index reflects the degree to which alternative i is worse than alternative i'. Analogous to the concordance index a discordance set $D_{ii'}$ is defined: this is the set of effects for which alternative i is worse than alternative i'. For each effect from this set the difference between the standardized scores of both alternatives is calculated. The discordance index is defined as the largest of these differences. This reflects the idea that, beyond a certain level, bad performance on one effect cannot be compensated for by good performance on the other effect.

Thresholds to be supplied by the decision maker, in combination with the concordance and discordance tables, are used to establish a weak and a strong outranking relationship between each pair of alternatives. A procedure of step-by-step elimination is used to transform the weak and the strong graph representing these outranking relationships into an overall ranking of the alternatives.

Your input and DEFINITE's output

Enter:

1. **Weights** (if not yet available);
2. **Standardization** (if changes to the default standardization are required);
3. **<F1>** to **execute** the method;
4. Strong and weak threshold values for the concordance table;
5. Strong and weak threshold values for the discordance table.

This results in:

1. A concordance table and a discordance table;
2. Strong and weak graphs before and after eliminating cycles;
3. One or more rankings of the alternatives.

Example

Effects table. Improve transportation to Nyverhoeck.

effects	units	Highway	Road/bus	Train
Costs	mln $	-200	-250	-400
Travel time	min/prs/day	30	15	10
Capacity	mln/km/yr	20	30	40
Emissions NO_x	ton/yr	-1000	-750	-100
Landscape	hectare	-10	-12	-2

Note that Travel time reflects the reduction in travel time (benefit) and Landscape reflects the loss of landscape (cost) that result from the alternatives.

Your input

Enter:

1. **Weights** (see also procedure C3):

Costs	0.2
Travel time	0.1
Capacity	0.3
Emissions NO_x	0.3
Landscape	0.1

2. Do not select standardization (default standardization is used).
3. Select **execute** to display the concordance and discordance tables. If intermediate results are presented you are asked to indicate at two levels (strong and weak graph) to what extent an alternative should be better (at least) and worse (at most) than another in order to achieve a high ranking.

DEFINITE's output

The first intermediate result is the concordance table below. The value 0.3 for the alternatives Highway and Train, for example, is calculated as the sum of the weights of the effects in which Highway scores better than Train or equal to Train: in this case the effects Costs with weight 0.2 and the effect Travel time with weight 0.1.

Concordance table. Improve transportation to Nyverhoeck.

	Highway	Road/bus	Train
Highway	-	0.40	0.30
Road/bus	0.60	-	0.30
Train	0.70	0.70	-

You are now asked to specify a strong and a weak threshold value for the concordance table. The value between brackets is calculated as 1.1 (strong value) and 0.9 (weak value) times the average of the concordance indices with a minimum value of 0.5. If presentation of intermediate results is switched off these values are assigned automatically.

Enter:

4. Strong concordance threshold: **0.6 <RETURN>**
   Weak concordance threshold: **0.5 <RETURN>**

The second intermediate result is the discordance table.

For example the value of 0.25 in the comparison between Highway and Road/bus is calculated as follows. Highway scores worse than Road/bus on the effects Capacity and Emissions NO_x. After standardization (see procedure C5) the difference for both effects equals 0.25.

Discordance table. Improve transportation to Nyverhoeck.

	Highway	Road/bus	Train
Highway	-	0.25	0.90
Road/bus	0.50	-	0.83
Train	0.67	0.38	-

Enter:

5. Strong discordance threshold: **0.50 <RETURN>**
   Weak discordance threshold: **0.75 <RETURN>**

The procedure translates this information into a strong and weak graph. The 1 in the comparison between Train and Road/bus indicates that Train is preferred to Road/bus.

Strong graph. Improve transportation to Nyverhoeck.

	Highway	Road/bus	Train
Highway	-	-	-
Road/bus	1	-	-
Train	-	1	-

Weak graph. Improve transportation to Nyverhoeck.

	Highway	Road/bus	Train
Highway	-	-	-
Road/bus	1	-	-
Train	1	1	-

One or more cycles such as, Alt.1 > Alt.2 **and** Alt.2 > Alt.1 may occur in both graphs. If a cycle occurs all alternatives within this cycle are combined into a group of alternatives with the same rank (see also under comments). In this example no cycles are found. The strong graph/cycles and the weak graph/cycles are therefore identical to the strong and weak graph.

Both graphs are shown in Figure 6.2. The solid arrows represent the strong outranking relationships and the dotted arrow the weak outranking relationship.

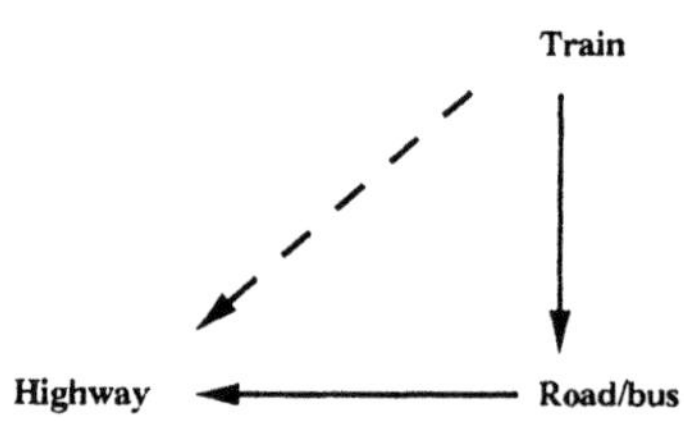

Figure 6.2. An outranking graph of the Nyverhoeck example.

A procedure of step-by-step elimination is used to transform the weak and strong graph to an overall ranking of the alternatives. First all alternatives that have no incoming solid arrows are selected. In this example this is only the case for the Train alternative. If two or more alternatives have no incoming solid arrows they are ordered according to their dotted arrows.

The selected alternatives are removed from the graph and the procedure is repeated. This results in the following first order ranking: Train > Road/bus > Highway.

All arrows are then reversed and the procedure is repeated. After reversing the resulting ranking (A > B is changed to B > A) this results in the following second order ranking: Train > Road/bus > Highway. The final ranking is determined as the average of the first and second order ranking.

Results of the ELECTRE method. Improve transportation to Nyverhoeck.

weights rank	score	ranking rank
1. Capacity	0.300	1. Train
Emissions NO_x	0.300	2. Road/bus
3. Costs	0.200	3. Highway
4. Travel time	0.100	
Landscape	0.100	

Comments

Type of results

A procedure of step-by-step elimination is used to transform the weak and strong graph representing the weak and strong outranking relationships into an overall ranking of the alternatives. The method does not always result in a complete ranking and apart from the ranking no evaluation scores are provided. In some cases Electre produces two or more partial rankings. In case of two partial rankings DEFINITE displays both rankings. In case only one alternative cannot be ranked to all other alternatives this alternative is displayed with a ? If more than two partial rankings are produced the two rankings that include the most alternatives are shown.

Cycles

Cycles can occur in the strong or weak graph. A cycle can include two alternatives, for example, Alt.1 > Alt.2 **and** Alt.2 > Alt.1 but also three or more alternatives, for example, Alt.1 > Alt.2 **and** Alt.2 > Alt.3 **and** Alt.3 > Alt.1. If a cycle occurs all alternatives within this cycle are combined into a group of alternatives with the same rank. To produce the graph shown below the values of the strong thresholds are changed to 0.5 (concordance) and 0.5 (discordance). As can be seen from the table Highway > Road/bus **and** Road/bus > Highway.

Strong graph (0.5, 0.5) Improve transportation to Nyverhoeck.

	Highway	Road/bus	Train
Highway	-	1	-
Road/bus	1	-	-
Train	1	1	-

In the next step alternatives Highway and Road/bus are combined into the group (Highway = Road/bus). In the strong graph/without cycles DEFINITE only shows the values for the first alternative in a group and presents the other alternatives as an empty column. In this example Highway and Road/bus are combined. The 1 in the comparison between Train and Highway applies therefore to both Highway and Road/bus. Road/bus is shown as an empty column. If cycles occur it is recommended to translate the table representation of the graph into a arrow representation of the graph (see Figure 6.2).

Strong graph/cycles. Improve transportation to Nyverhoeck.

	Highway	Road/bus	Train
Highway	-	-	-
Road/bus	-	-	-
Train	1	-	-

Saving the result

Scores and ranking are dependent on the standardization method applied and the threshold values. The results of different calculations can be compared by saving the MCA results. In this way the influence of changes in weights, weight methods, and standardization procedures can be analyzed.

It is recommended to save each relevant result of a multicriteria method. A result that is saved can be analyzed using sensitivity analysis in module D and will be included in the report generated in module E. When you save a result you are asked whether you want to overwrite the previous result. This question is asked to prevent saving irrelevant results. Answer **no** if you want to save both results.

To be able to identify the various results to be analyzed in module D each result receives a name and information line. For example, **electre.001** identifies the first result of the Electre method that was saved.

The information line can be used to further identify the result. As default the information line provides the weight method used and the value of the highest weight. This information line can be replaced or extended. The information line: 'Direct: capacity = 0.400, reference set' indicates that this result was obtained using direct assessment of weights, capacity has weight 4.00 and the weight set was labelled 'reference set'.

Standardization

If procedure C5, **standardization**, is *not* selected each row of effect scores is automatically standardized as follows:

- monetary, ratio, and interval scale : score / highest score
- ordinal scale : reverse order followed by score / highest score
- binary scale : yes = 1; no = 0 *or* yes = 0; no = 1;
- ---/+++ scale : ---, --, -, 0, +, ++, +++ = 0, 1/6, 2/6, 3/6, 4/6, 5/6, 1.

PROCEDURE C2.3. REGIME METHOD

Purpose of this procedure

This procedure uses a random generator to estimate for each pair of alternatives the probability that one alternative ranks above another alternative. If you have not yet assigned any weights, you need to do so within this procedure. The Regime method uses qualitative weights only.

Description

The Regime method can be viewed as an ordinal generalization of pairwise comparison methods such as the Electre method. The starting point of the Regime method is the concordance index $c_{ii'}$. The focus of this method is on the sign of $c_{ii'}$ - $c_{i'i}$ for each pair of alternatives. If this sign is positive, alternative i is preferred to i'; and the reverse holds if the sign is negative.

Ordinal weights are interpreted as unknown quantitative weights. A set S is defined containing all sets of quantitative weights that conform to the qualitative priority information. In some cases the sign will be the same for the whole set S and the alternatives can be ranked accordingly. In other cases the sign of the pairwise comparison cannot be determined unambiguously: for parts of the set S the sign of $c_{ii'}$-$c_{i'i}$ is positive and for other parts it is negative. The distribution of the weights within S is assumed to be uniform and therefore the relative sizes of the subsets of S can be interpreted as the probability that alternative i is preferred to alternative i'. Probabilities are aggregated to produce an overall ranking of the alternatives. The relative sizes of the subsets are estimated using a random generator.

Two versions of the Regime method are available: a version designed for evaluation problems with only qualitative information on scores and a version designed for evaluation problems with a combination of qualitative and quantitative information on scores. DEFINITE includes both versions and will select one of the two versions automatically according to the type of information that is available.

The concept of qualitative scores as unknown quantitative scores and the use of a random generator to generate quantitative representations of the qualitative scores makes it possible to treat quantitative and qualitative scores in the same way. The Regime method can therefore be applied to mixed problems without losing the information contained in the quantitative scores.

Your input and DEFINITE's output

Enter:

1. Qualitative **weights** (if not yet available);
2. **<F1>** to execute the method.

This results in:

1. A probability table;
2. A ranking of the alternatives;
3. An appraisal score for each alternative.

Example

In this example it is assumed that only ordinal information is available for two of the effects. It is also assumed that only ordinal information is available on weights.

Effects table. Improve transportation to Nyverhoeck.

effects	units	Highway	Road/bus	Train
Costs	mln $	-200	-250	-400
Travel time	---/+++	+++	++	+
Capacity	mln/km/yr	20	30	40
Emissions NO_x	ton/yr	-1000	-750	-100
Landscape	---/+++	---	---	-

Your input

Enter:

1. **Weights** (see also procedure C3):
   1. Capacity
      Emissions NO_x
   3. Costs
   4. Travel time
      Landscape
2. **Execute.**

Note that in this case it makes no sense to use the Expected Value method, the Extreme Value method, or the Random Value method. The Pairwise Comparison method, in this context, is only used for deducing qualitative weights.

DEFINITE's output

After selecting **execute** the probability table is displayed. The table below shows the relative success indices for the transportation example. The value 1.00 in the comparison between Train and Highway alternatives indicates that, for all quantitative values that comply with the qualitative scores of the effects Travel time and Landscape, the Train alternative ranks above the Highway alternative. For this comparison, therefore, no added value is to be expected from measurement of these effects on a higher measurement scale. The probability that, given the ordinal information on Travel time and Landscape, the Road/bus alternative ranks higher than the Highway alternative equals 92%.

Probability table. Improve transportation to Nyverhoeck.

	Highway	Road/bus	Train
Highway	-	0.08	0.00
Road/bus	0.92	-	0.00
Train	1.00	1.00	-

The overall score of an alternative is calculated as the row average of the relative success indices. In this example Train is the most preferred alternative, followed by the Road/bus and Highway. The following table shows the weights used, the ranking and the scores.

Results of the Regime method. Improve transportation to Nyverhoeck.

weights rank		ranking rank		score
1.	Capacity	1.	Train	1.00
	Emissions NO_x	2.	Road/bus	0.46
3.	Costs	3.	Highway	0.04
4.	Travel time			
	Landscape			

Comments

Calculation time

The Regime method calculates the ranking of the alternatives a large number of times using a random generator. On slow computers this can cause some waiting time. The **<ESC>** key can be used to interrupt the calculations. Interrupting the calculations may lead to inaccurate results.

Weights

Since the method uses only ordinal information on weights, the weights methods which are specially assigned to generate quantitative weights are not available.

Saving the result

It is recommended to save each relevant result of a multicriteria method. A result that is saved can be analyzed using sensitivity analysis in module D and will be included in the report generated in module E. When you save a result you are asked whether you want to overwrite the previous result. This question is asked to prevent saving irrelevant results. Answer **no** if you want to save both results.

To be able to identify the various results to be analyzed in module D each result receives a name and information line. For example, **regime.001** identifies the first result of the Regime method that was saved.

The information line can be used to further identify the result. As default the information line provides the weight method used and the value of the highest weight. This information line can be replaced or extended. The information line: 'Ordinal, capacity, reference set' indicates that this result was obtained using ordinal weights, that capacity is the most important effect and that the weight set was labelled 'reference set'.

Standardization

The Regime method does not standardize qualitative (ordinal and ---/+++) effects. Do not standardize ordinal effects with procedure C5 when applying the Regime method because this leads to incorrect results. If procedure C5, **standardization,** is *not* selected each row of effect scores is automatically standardized as follows:

- monetary, ratio,
  interval and
  ordinal scale : no standardization
- binary scale : yes = 1; no = 0 *or* yes = 0; no = 1;
- ---/+++ scale : no standardization

PROCEDURE C2.4. EXPECTED VALUE METHOD

Purpose of this procedure

This procedure generates a ranking of the alternatives.

Description

As a first step all effect scores are standardized according to the Expected Value method. An appraisal score is then calculated for each alternative by first multiplying these standardized effect scores by its appropriate weight followed by summing the weighted scores of all effects.

Your input and DEFINITE's output

Enter:

1. **Weights** (if not yet available);
2. **Standardization** (if changes to the default standardization are required;
3. **<F1>** to execute the method.

This results in:

1. An effects table standardized according to the Expected Value method;
2. A ranking of the alternatives;
3. An appraisal score for each alternative.

Example

Effects table. Improve transportation to Nyverhoeck.

effects	units	Highway	Road/bus	Train
Costs	mln $	-200	-250	-400
Travel time	---/+++	+++	++	+
Capacity	mln/km/yr	20	30	40
Emissions NO_x	ton/yr	-1000	-750	-100
Landscape	---/+++	---	---	-

Your input

Enter:

1. **Weights** (see also procedure C3):
   1. Capacity
      Emissions NO_x
   2. Costs
   3. Travel time
      Landscape
2. Do not select standardization in order to use default standardization;
3. **Execute.**

DEFINITE's output

An intermediate result is the effects table below. The qualitative scores of Travel time and Landscape are standardized with the Expected Value method (all others according to the default standardization shown below).

Standardized effects table. Improve transportation to Nyverhoeck.

effects	units	Highway	Road/bus	Train
Costs	mln $	1.00	0.75	-
Travel time	min/prs/day	1.00	0.89	0.61
Capacity	mln/km/yr	-	0.50	1.00
Emissions NO_x	ton/yr	-	0.28	1.00
Landscape	hectare	0.75	0.75	1.00

The final result is the table below. This table shows the weights, the ranking and the appraisal scores.

Results of the Expected Value method. Improve transportation to Nyverhoeck.

weights rank	score	ranking rank	score
1. Capacity	0.357	1. Train	0.82
Emissions NO_x	0.357	2. Road/bus	0.50
3. Costs	0.157	3. Highway	0.27
4. Travel time	0.065		
Landscape	0.065		

Comments

Saving the result

It is recommended to save each relevant result of a multicriteria method. A result that is saved can be analyzed using sensitivity analysis in module D and will be included in the report generated in module E. When you save a result you are asked whether you want to overwrite the previous result. This question is asked to prevent saving irrelevant results. Answer **no** if you want to save both results.

To be able to identify the various results to be analyzed in module D each result receives a name and information line. For example, **exp.001** identifies the first result of the Expected Value method that was saved.

The information line can be used to further identify the result. As default the information line provides the weight method used and the value of the highest weight. This information line can be replaced or extended. The information line: 'Direct: Capacity = 0.400, reference set' indicates that this result was obtained using direct assessment of weights, Capacity has weight 4.00 and the weight set was labelled 'reference set'.

Standardization

Scores and ranking are dependent on the standardization method chosen. The results of different calculations can be compared by saving the MCA results. In this way the influence of changes in weights, weights methods and standardization procedures can be analyzed.

If procedure C5, **standardization**, is *not* selected each row of effect scores is automatically standardized as follows:

- monetary, ratio and interval scale : (score-lowest score) / (highest score-lowest score)
- ordinal scale : expected value
- binary scale : yes = 1; no = 0 *or* yes = 0; no = 1;
- ---/+++ scale : expected value

PROCEDURE C2.5. EVAMIX METHOD

Purpose of this procedure

This procedure generates a ranking of the alternatives. The Evamix method generates separate dominance tables for quantitative and for qualitative effect scores. Both tables are combined into a total dominance table.

Description

The Evamix method is designed to deal with an effects table with a mixture of qualitative and quantitative effects. The set of effects in the effects table is divided into a set of ordinal effects *O* and a set of quantitative effects *Q*. Dominance effects are calculated for both sets.

The method requires quantitative weights but can be used in combination with any of the methods dealing with ordinal priority information (procedure C3). A total dominance index is found by combining the indices that are calculated separately for the qualitative and quantitative scores. These indices need to be standardized before they can be combined. The most straightforward standardization divides qualitative indices by the absolute value of their sum and does the same with quantitative indices. Next the qualitative dominance indices are multiplied by the sum of the qualitative weights and the quantitative dominance indices are multiplied by the sum of the quantitative weights. The total dominance indices are calculated as the weighted sum of the qualitative and quantitative dominance indices. The final scores for the alternatives are found as the row totals of the total dominance table.

Your input and DEFINITE's output

Enter:

1. **Weights** (if not yet available);
2. **Standardization** (if changes to the default standardization are required;
3. **<F1>** to **execute** the method.

This results in:

1. A qualitative and a quantitative dominance table;
2. A total dominance table;
3. A ranking of the alternatives;
4. An appraisal score for each alternative.

Example

Effects table. Improve transportation to Nyverhoeck.

effects	units	Highway	Road/bus	Train
Costs	mln $	-200	-250	-400
Travel time	---/+++	+++	++	+
Capacity	mln/km/yr	20	30	40
Emissions NO_x	ton/yr	-1000	-750	-100
Landscape	---/+++	---	---	-

Your input

Enter:

1. **Weights** (see also procedure C3):

Costs	0.2
Travel time	0.1
Capacity	0.3
Emissions NO_x	0.3
Landscape	0.1

2. Do not select standardization in order to use default standardization;
3. **Execute**.

DEFINITE's output

Qualitative dominance table. Improve transportation to Nyverhoeck.

	Highway	Road/bus	Train
Highway	-	0.10	-
Road/bus	-0.10	-	-
Train	-	-	-

Quantitative dominance table. Improve transportation to Nyverhoeck.

	Highway	Road/bus	Train
Highway	-	-0.18	-0.40
Road/bus	0.18	-	-0.22
Train	0.40	0.22	-

Total dominance table. Improve transportation to Nyverhoeck.

	Highway	Road/bus	Train
Highway	-	8.3E-03	-0.20
Road/bus	-8.3E-03	-	-0.11
Train	0.20	0.11	-

The table below is the final result. This table shows the weights, the ranking and the appraisal scores.

Results of the Evamix method. Improve transportation to Nyverhoeck.

weights rank		score	ranking rank	score
1.	Capacity	0.300	1. Train	0.31
	Emissions NO_x	0.300	2. Road/bus	-0.12
3.	Costs	0.200	3. Highway	-0.19
4.	Travel time	0.100		
	Landscape	0.100		

Comments

Saving the result

It is recommended to save each relevant result of a multicriteria method. A result that is saved can be analyzed using sensitivity analysis in module D and will be included in the report generated in module E. When you save a result you are asked whether you want to overwrite the previous result. This question is asked to prevent saving irrelevant results. Answer **no** if you want to save both results.

To be able to identify the various results to be analyzed in module D each result receives a name and information line. For example, **evamix.001** identifies the first result of the Evamix method that was saved.

The information line can be used to further identify the result. As default the information line provides the weight method used and the value of the highest weight. This information line can be replaced or extended. The information line: 'Direct: Capacity = 0.400, reference set' indicates that this result was obtained using direct assessment of weights, Capacity has weight 4.00 and the weight set was labelled 'reference set'.

Standardization

Scores and ranking are dependent on the standardization method and weights. Save the MCA results to compare the results of different calculations. In this way the influence of changes in weights, weighting methods and standardization procedure can be analyzed.

The Evamix method does not standardize qualitative (ordinal and ---/+++) effects. Do not standardize ordinal effects with procedure C5 when applying the Evamix method because this leads to incorrect results. If procedure C5, **standardization**, is *not* selected each row of effect scores is automatically standardized as follows:

- monetary, ratio and
  interval scale: (score-lowest score) / (highest score-lowest score)
- ordinal scale : no standardization
- binary scale : yes = 1; no = 0 *or* yes = 0; no = 1;
- ---/+++ scale : no standardization

PROCEDURE C3. ASSIGN WEIGHTS

Purpose of this procedure

The use of multicriteria methods requires information on the relative importance of each effect: you need to assign weights. In the Regime method it is sufficient to order the effects from most to least important. All other methods require quantitative weights.

Description

In this procedure you can directly enter quantitative weights (direct assessment) or use methods that transform qualitative priority statements into quantitative weights (Expected Value method, Pairwise Comparison method, Random Value method and the Extreme Value method). The weights determined in this procedure can be saved and used in each of the multicriteria methods. The Random Value and Extreme Value methods directly produce the result of the selected multicriteria method. The weights assigned in this procedure can be saved and used in each of the multicriteria methods.

The methods available are:

1. **Expected Value method**
   In the Expected Value method the user is asked to arrange the effects from most important to least important. The procedure uses the expected value within this ordering to calculate quantitative weights.

2. **Direct assessment**

3. **Pairwise Comparison method**
   This method is also known as the Analytical Hierarchy Process (AHP). The user is asked to select the most important of each pair of effects. Subsequently the user is asked in qualitative terms to what extent the first effect is more important than the other. The method converts these comparisons of all pairs of effects to quantitative weights for all effects.

4. **Random Value method**
   In the Random Weights method the user is asked to arrange the effects from most important to least important. The weights will be found by random drawings of weights which fulfil two conditions:
   1. The value of the weights must correspond to the ranking entered by the user.

2. The sum of weights of all effects is one. The ranking of the alternatives is determined for 500 sets of weights generated within these conditions by a random generator. These rankings will be converted into a frequency table and further converted into one ranking of the alternatives.

5. **Extreme Value method**
   In the Extreme Weights method the user is asked to arrange the effects from most important to least important. This method is based on the extreme quantitative weights which just fulfil the assigned qualitative ranking of the weights. Because there are always different extreme combinations of weights, this method can result in different rankings of alternatives. In this cases it is impossible to determine a complete ranking of the alternatives.

Your input and DEFINITE's output

Enter:

1. The **weight** method you want to use;
2. **Make** to make a new set of weights;
3. The requested priority information.

This results in:

A set of weights.

or

An MCA result (if the Random Value or Extreme Value method is used).

Example

Starting point in this example are the effects:

Costs, Travel time, Capacity, Emissions NO_x and Landscape.

Your input

Enter:

1. **Expected Value** method, **Random Value** method or **Extreme Value** method;
2. **Make**;
3. The priority ranking of the effects. In this example: Capacity is as important as Emissions NO_x. Emissions NO_x are more important than Costs. Costs are more important than Travel time. And, finally, Travel time is as important as Landscape.

or

1. **Direct assessment**;
2. **Make**;
3. A quantitative weight for each effect. For example:

Capacity: 50, Emissions NO_x: 50, Costs: 40, Travel time 30 and Landscape also 30.

or

1. **Pairwise Comparison method;**
2. **Make;**
3. An assessment for every pair of effects. For example:
   Capacity is equally important as Emissions NO_x. Capacity is more important than Costs. Capacity is strongly more important than Travel time, etc..

DEFINITE's output

Expected Value method

1.	Capacity	0.357
	Emissions NO_x	0.357
3.	Costs	0.157
4.	Travel time	0.065
	Landscape	0.065

Direct assessment

1.	Capacity	0.250
	Emissions NO_x	0.250
3.	Costs	0.200
4.	Travel time	0.150
	Landscape	0.150

Pairwise Comparison method

As an intermediate result the following table of pairwise comparisons is presented. Note that if, for example, Costs are considered more important than Travel time, receiving a score 3, the reverse comparison of Travel time against Costs automatically receives a score of 1/3 (0.33).

Table of pairwise comparisons. Improve transportation to Nyverhoeck.

	Costs	Travel time	Capacity	Emissions NO_x	Landscape
Costs	1.00	3.00	0.33	0.33	3.00
Travel time	0.33	1.00	0.20	0.20	1.00
Capacity	3.00	5.00	1.00	1.00	5.00
Emissions NO_x	3.00	5.00	1.00	1.00	5.00
Landscape	0.33	1.00	0.20	0.20	1.00

inconsistency ratio = 0.011.

From this table the following weights are calculated:

1.	Capacity	0.359
	Emissions NO_x	0.359
3.	Costs	0.153
4.	Travel time	0.064
	Landscape	0.064

Random Value method

This weighting method is used in this example in combination with Weighted Summation (see procedure C2.1). As an intermediate result the following frequency table is presented.

Frequency table. Improve transportation to Nyverhoeck.

	Rank 1	Rank 2	Rank 3
Highway		65	435
Road/bus		462	38
Train	500		

This table shows that, given the qualitaive ordering of weights, Train ranks as the best alternative for all quantitative weights that conform to the qualitative ordering (maximum = 500). The sum of Ranks 1 and 2 is highest for Road/bus (0 + 462); Road/bus is therefore the next best alternative, followed by Highway.

This results in the following ranking of alternatives:

1. Train
2. Road/bus
3. Highway

Extreme Value method

This method is used in this example in combination with Weighted Summation (see procedure C2.1).

The Extreme Value method determines a ranking of the alternatives for each extreme combination of weights that meet the qualitative ordering of weights. These rankings are shown in the following intermediate result.

Rankings in the extreme points. Improve transportation to Nyverhoeck.

Extreme points	Highway	Road/bus	Train
corner 01	3	2	1
corner 02	3	2	1
corner 03	2	3	1

This result shows that in all extreme points Train is the best alternative. Road/bus is better than Highway in only two out of three extreme points. The rank order of Road/bus and Highway can therefore not be determined. This results in the ranking shown below. Note that in this case the shared position of alternatives Road/bus and Highway does not imply that both alternatives have the same rank. No more can be concluded then that alternative Train is preferred to these alternatives.

1. Train
2. Road/bus
   Highway

Comments

Choice of a method

The choice of a weighting method is dependent on the choice of a multicriteria method.

The Expected Value weights differ from the Pairwise Comparison weights. In general, the Pairwise Comparison weights will better reflect the priorities of a decision maker. A disadvantage of the Pairwise Comparison method is the large number of comparisons to be made if more than seven effects are to be compared.

The Random Value and Extreme Value methods do not result in weights but are directly combined with a multicriteria method to produce a ranking of alternatives. The Random Value method always results in a complete ranking. Since this ranking may not conform to all possible quantitative weights, this ranking may not be completely certain. The probability table that is shown as an intermediate result provides an indication of this uncertainty. The Extreme Value method does not always lead to a complete ranking. The ranking that is provided, however, conforms to all possible quantitative weights and is therefore completely certain. All possible rankings are shown as an intermediate result.

The Regime method requires qualitative weights only. Methods to assess quantitative weights are therefore not available.

Direct assessment

Weights entered using direct assessment are divided by the sum of all weights to make these weights add up to one.

Pairwise comparison

The number of comparisons increases combinatorially with the number of effects (n*(n-1)/2). For example, the assessment of the weights of 4 effects involves no more than 6 comparisons, but assessment of the weights for 10 effects requires as many as 45 comparisons. It is not recommended to use this method if the number of effects is higher than 7 (21 comparisons). An inconsistency ratio is presented.[1]

---

1 See: Saaty, T.L., and L.G. Vargas (1984). Comparison of Eigenvalue and logarithmic least squares and least squares methods in estimating ratios. *Mathematical modelling*, Vol. 5, pp. 309-324.

PROCEDURE C4. PRESENTATION OF RESULTS

Purpose of this procedure

This procedure provides a graphic presentation of evaluation results.

Example

The result of the Expected Value method

Results of the Expected Value method. Improve transportation to Nyverhoeck.

weights rank		score	ranking rank	score
1.	Capacity	0.357	1. Train	0.82
	Emissions NO_x	0.357	2. Road/bus	0.50
3.	Costs	0.157	3. Highway	0.27
4.	Travel time	0.065		
	Landscape	0.065		

DEFINITE's output

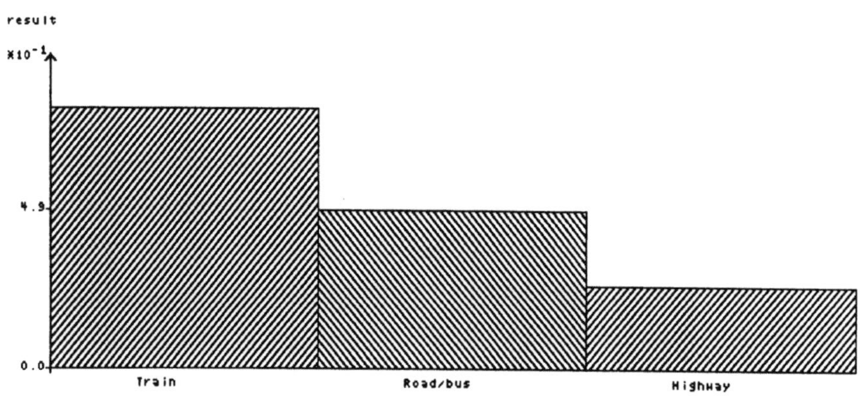

Figure 6.3. A graphic presentation of the results of the Expected Value method.

Comments

Printing the result

Figures are *not* included in the report that can be generated in module E. It is therefore advisable to print this figure now. Procedures to save screen output are available for including figures in a text file. Users of WordPerfect 5.0 or higher, for example, can use the GRAB program to include this figure in a WordPerfect document. See the WordPerfect manual for further instructions.

PROCEDURE C5. ADJUST STANDARDIZATIONS

Purpose of this procedure

Before a multicriteria method can be applied, for most methods, the effects table needs to be standardized. In DEFINITE, for each method the most appropriate method of standardization is applied by default.

If the default standardization is considered inappropriate this procedure can be used to select a specific method of standardization for the effects table as a whole, or for each effect individually.

First, the effects table is displayed; in the menu you need to indicate whether you want to standardize one effect, a category of effects or the whole effects table simultaneously. Next, select one of the available methods for standardization.

In this procedure five standardization methods are available:

1. **Division by row maximum**. The standardized scores receive values between -1 and 1.
2. **Interval standardization**. Scores will be standardized per effect with the following formula: (score - lowest score)/(highest score - lowest score). The standardized scores receive values between 0 and 1.
3. **Division by row total**. The standardized scores receive values between -1 and 1.
4. **Standardization by objectives**. You need to specify for every effect an ideal or objective value and a minimum value. A meaningful minimum value can be the score in the no action alternative or the score of the worst possible alternative.
5. **Vector normalization**. The scores per effect will be divided by a norm. This norm is equal to the square root of the sum of the squares of the scores of an effect. The standardized scores receive values between -1 and 1.

Your input and DEFINITE's output

Enter:

1. The **effect** or **category of effects** you want to standardize;
2. The **standardization method** to be used.

This results in:

A standardized effects table.

If you select **continue** the scores of the remaining effects are standardized automatically and you return to the multicriteria method you started with. At this point you have the option to save the specifications of the standardization you have applied. These specifications can be retrieved after reentering this procedure.

Example

Effects table. Improve transportation to Nyverhoeck.

effects	units	Highway	Road/bus	Train
Costs	mln $	-200	-250	-400
Travel time	min/prs/day	30	15	10
Capacity	mln/km/yr	20	30	40
Emissions NO_x	ton/yr	-1000	-750	-100
Landscape	hectare	-10	-12	-2

Note that Travel time reflects the reduction in travel time (benefit) and Landscape reflects the loss of landscape (cost) that result from the alternatives.

Division by row maximum

Your input

Enter:

1. **Row by row** or a **category of effects** to standardize. In this example the quantitative effects;
2. The standardization method. In this example **division by row maximum**.

DEFINITE's output

Standardized effects table. Improve transportation to Nyverhoeck.

effects	units	Highway	Road/bus	Train
Costs	mln $	-0.50	-0.63	-1.00
Travel time	min/prs/day	1.00	0.50	0.33
Capacity	mln/km/yr	0.50	0.75	1.00
Emissions NO_x	ton/yr	-1.00	-0.75	-0.10
Landscape	hectare	-0.83	-1.00	-0.17

Interval standardization

Your input

Enter:

1. **Row by row** or a **category of effects** to standardize. In this example the quantitative effects;
2. The standardization method. In this example **interval standardization**.

DEFINITE's output

Standardized effects table. Improve transportation to Nyverhoeck.

effects	units	Highway	Road/bus	Train
Costs	mln $	1.00	0.75	-
Travel time	min/prs/day	1.00	0.25	-
Capacity	mln/km/yr	-	0.50	1.00
Emissions NO_x	ton/yr	-	0.28	1.00
Landscape	hectare	0.20	-	1.00

Division by row total

Your input

Enter:

1. **Row by row** or a **category of effects** to standardize. In this example the quantitative effects;
2. The standardization method. In this example **division by row total.**

DEFINITE's output

Standardized effects table. Improve transportation to Nyverhoeck.

effects	units	Highway	Road/bus	Train
Costs	mln $	-0.24	-0.29	-0.47
Travel time	min/prs/day	0.55	0.27	0.18
Capacity	mln/km/yr	0.22	0.33	0.44
Emissions NO_x	ton/yr	-0.54	-0.41	-0.05
Landscape	hectare	-0.42	-0.50	-0.08

Standardization by objectives

Your input

Enter:

1. **Row by row** and select the **effects**. In this example **Travel time** and **Emissions NO_x**;
2. The standardization method. In this example **standardization by objectives.**

3. The **objective** and **minimum values**. In this example for Travel time an objective of 100 minutes and a minimum value of 0 are specified. For Emissions NO_x the objective is 0 Emissions NO_x and the minimum value equals 1000 ton/year.

DEFINITE's output

Standardized effects table. Improve transportation to Nyverhoeck.

effects	units	Highway	Road/bus	Train
Costs	mln $			
Travel time	min/prs/day	0.30	0.15	0.10
Capacity	mln/km/yr			
Emissions NO_x	ton/yr	0	0.25	0.90
Landscape	hectare			

The standardized score equals:

$$\frac{(\text{score} - \text{minimum value})}{(\text{objective} - \text{minimum value})}.$$

Vector normalization

Your input

Enter:

1. **Row by row** or a **category of effects** to standardize. In this example the quantitative effects;
2. The standardization method. In this example **vector normalization**.

DEFINITE's output

Standardized effects table. Improve transportation to Nyverhoeck.

effects	units	Highway	Road/bus	Train
Costs	mln $	-0.39	-0.49	-0.78
Travel time	min/prs/day	0.86	0.43	0.29
Capacity	mln/km/yr	0.37	0.56	0.74
Emissions NO_x	ton/yr	-0.80	-0.60	-0.08
Landscape	hectare	-0.64	-0.76	-0.13

PROCEDURE C6.1. COST-BENEFIT ANALYSIS

Purpose of this procedure

This procedure calculates the net present value, the benefit-cost ratio and the internal rate of return for each of the alternatives. This procedure uses the cost-benefit table created in module B. The internal rate of return can only be calculated if effects tables have been created for more than one year.

Description

Three decision rules are generally used in CBA:
1. The net present value (NPV) of an alternative, defined as the balance of discounted benefits and discounted costs;
2. The benefit-cost ratio (BCR) of an alternative, defined as the ratio of discounted benefits to discounted costs; and
3. The internal rate of return (IRR) of an alternative, defined as the discount rate that results in equality of discounted benefits and discounted costs.

The choice of a decision rule depends on the decision problem. The NPV is most appropriate for problems when one alternative must be selected from mutually exclusive alternatives and no other investment opportunities are available. The BCR and IRR are appropriate if alternatives can be combined and other investment opportunities are available. If costs do not set constraints on the decision all three rules can be used to decide whether an alternative should or should not be implemented.

The result of the procedure is a table which displays the net present value, the benefit-cost ratio and the internal rate of return for each alternative. A ranking of the alternatives is found by selecting **sort NPV**, **sort B/C** or **sort IRR**.

Your input and DEFINITE's output

Enter:

No additional input is required.

This results in:

1. The cost-benefit table defined in the previous module;
2. The balances of benefits and costs for each year (if relevant);
3. The net present values, benefit-cost ratios and internal rates of return (if available) of the alternatives.

Select **'graph NPV'**, **'graph B/C'** or **'graph IRR'** to obtain a graphic presentation of the net present values, the cost-benefit ratios or the internal rates of return.

Example

Monetary cost-benefit table (scores in millions). Improve transportation to Nyverhoeck.

effects	Cost Benefit	units	prices millions	Highway	Road/bus	Train
Costs	C	millions	1	218.18	272.73	436.36
Emissions NO_x	C	ton/yr	?	p.m.	p.m.	p.m.
Landscape	C	hectare	10	100.00	120.00	20.00
Travel time	B	min/prs/day	4	229.08	114.56	76.36
Capacity	B	mln/km/yr	?	p.m.	p.m.	p.m.

Your input

Enter:

This procedure does not need any new input.

DEFINITE's output

As an intermediate result the balances of cost and benefits for all specified years are presented. As the final result the net present value, the benefit-cost ratio and the internal rate of return are presented.

Balance of costs and benefits per year. Improve transportation to Nyverhoeck.

	Highway	Road/bus	Train
1988	-180	-310	-380
1989	100	35	

Results of Cost-Benefit analysis (scores in millions). Improve transportation to Nyverhoeck.

alternative discount rate: 5.00%	net present value	costs/ benefits	internal rate of return (%)
1. Highway	-89.09	0.72	negative
2. Road/bus	-278.18	0.29	negative
3. Train	-380.00	0.19	unknown

Comments

Interpretation of the results

The rankings determined using the net present values, the benefit-cost ratios and the internal rates of return may not be identical. Each of the three indices is based on different assumptions. The net present value is most appropriate if one alternative must be selected from mutually exclusive alternatives and no other investment opportunities are available. The benefit-cost ratio and internal rate of return are appropriate if alternatives can be combined and other alternatives are available.

Graphical presentations

Graphical presentations can be made of the net present values, the benefit-cost ratios and the internal rates of return.

Figures are *not* included in the report that can be generated in module E. It is therefore advisable to print this figure now. Procedures to save screen output are available for including figures in a text file. Users of WordPerfect 5.0 or up, for example, can use the GRAB programme to include this figure in a WordPerfect document. See the WordPerfect manual for further instructions.

Calculation of the discounted values

The standard effects table is the effects table in the base year (year 0). Costs and benefits that occur after the base year are included in time-effect tables (time.***). These time-effect tables are linked to specific years. A maximum of 50 time-effect tables can be entered.

Costs and benefits that occur during a certain year are attributed to the end of that year. Figure 6.4 shows that costs and benefits that occur in year 2 are attributed to the end of year 2 before being discounted over 2 years to the base year. Costs and benefits occurring in the base year are not discounted. This implies that the investment decision is assumed to take place at the first of January of year 1. It is assumed implicitly that costs and benefits in year 0 occur at the end of this year. Costs in year 0 are usually investment costs and there are usually no benefits. If the investment costs are spread over a longer period it can be preferable to also discount the investment costs. The year of investment should then be entered as year 1 and the base year as the year before. In this case the effects table of the base year should be empty.

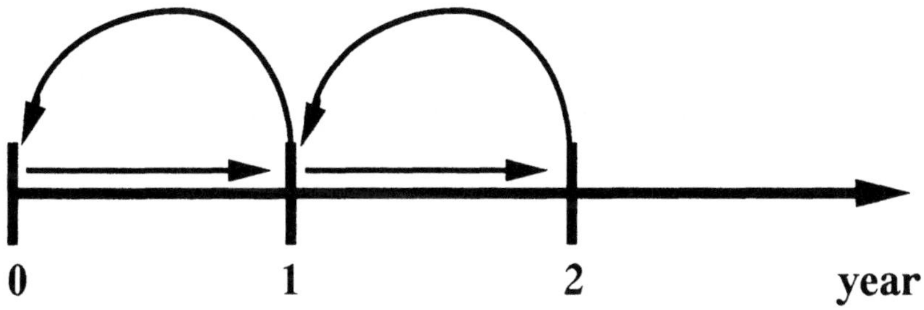

Figure 6.4. Effects tables, time and discounting.

The equation used to discount future costs and benefits is shown below.

$$x_{ji}^{d} = \sum_{t^0}^{t^n} x_{ji}^{t} * \frac{(1+p_i/100)^{t-t^0}}{(1+r_i/100)^{t-t^0}}$$

In this equation x_{ji}^{d} represents the discounted effect i of alternative j and x_{ji}^{t} represents this effect at year t. The base year is shown as t^0 and the last year to be included (the planning horizon) as t^n. In this equation r_i represents the discount rate and as p_i the price index, both entered as percentages. The price index can be used to correct for changes in prices over time. The discount rate can be specified as a uniform discount rate for all effects or can be specified for each effect separately.

Calculation of the internal rate of return

In the following cases the internal rate of return can not be calculated or may produce irrelevant results if:

- all costs and benefits occur in the same year;
- costs and benefits are the same for each year;
- a project generates only costs or only benefits;
- benefits occur before costs;
- the sum of undiscounted future benefits is smaller than the sum of undiscounted costs.

Depending on the distribution of benefits and costs over time DEFINITE might either produce no internal rate of return (unknown), a negative rate of return (negative) or an extremely high rate of return (>100). If costs occur both in the beginning of the project and at the end of the project it is possible that two different internal rates of return exist. DEFINITE will in these cases only provide the smallest internal rate of return.

PROCEDURE C6.2. COST-EFFECTIVENESS ANALYSIS

Purpose of this procedure

This procedure can be used to rank the alternatives according to total costs, according to total benefits, or according to the cost-effectiveness in achieving a selected benefit.

Total costs

Your input and DEFINITE's output

Enter:

Cost min.

This results in:

A ranking of the alternatives according to total costs.

Total benefits

Your input and DEFINITE's output

Enter:

Effect max.

This results in:

A ranking of the alternatives according to total benefits.

Cost effectiveness

Your input and DEFINITE's output

Enter:

1. **Cost effective;**
2. A benefit category.

This results in:

A ranking of the alternatives according to a selected benefit divided by total costs.

Example

Starting point in this example is the costs-benefit table below.

Discounted cost-benefit table. Improve transportation to Nyverhoeck.

effects	Cost Benefit	units	prices millions	Highway	Road/bus	Train
Costs	C	millions	1	218.18	272.73	436.36
Emissions NO_x	C	ton/yr	?	1909.09	1431.82	190.91
Landscape	C	hectare	10	10.00	12.00	2.00
Travel time	B	min/prs/day	4	57.27	28.64	19.09
Capacity	B	mln/km/yr	?	38.18	57.27	76.36

Note that Travel time reflects the reduction in travel time (benefit) and Landscape reflects the loss of landscape (cost) that result from the alternatives.

Your input

Enter:

1. **Cost effective**;
2. A benefit category. In this example **Capacity**.

DEFINITE's output

The procedure now calculates total costs, total benefits and the ratio capacity and total costs. The alternatives are ranked according to their cost-effectiveness with respect to capacity.

Cost-Effectiveness analysis. Improve transportation to Nyverhoeck.

alternatives	total costs millions	total benefits millions	Capacity/tot.costs person/yr/10E6
1. Train	456.36	76.36	0.17
2. Road/bus	392.73	144.55	0.15
3. Highway	318.18	229.09	0.12

Comments

If cost-effective is *not* selected DEFINITE automatically selects the first benefit effect from the cost-benefit table.

7. SENSITIVITY ANALYSIS PROCEDURES

Introduction

The sensitivities of the rankings obtained in module C to changes in evaluation methods, effect scores and weights are analyzed in **module D: sensitivity analysis**. Sensitivity analysis is supported by the following procedures:

D1. **Method uncertainty** to assess the sensitivity of the ranking to the evaluation method applied.

D2. **Effects uncertainty** to assess the influence of uncertainties in scores on the ranking of the alternatives.

D3. **Weights uncertainty** to assess the influence of uncertainties in weights on the ranking of the alternatives.

D4. **Weight intervals** to determine the intervals within which the rank order of two alternatives is insensitive to changes in weights.

D5. **Effect intervals** to determine the intervals within which the rank order of two alternatives is insensitive to changes in scores.

D6. **Costs and benefits uncertainty** to assess the influence of uncertainties in costs and benefits scores on the results of Cost-Benefit analysis.

D7. **Costs and benefits intervals** to determine the intervals within which the rank order of two alternatives is insensitive to changes in costs and benefits.

D8. **Price intervals** to determine the intervals within which the rank order of two alternatives is insensitive to changes in prices used in Cost-Benefit analysis.

D9. **Problem analysis** analyzes the results of all results of sensitivity analysis within a session and provides advice on: additional analysis, presentation of the final results.

A flow diagram of the procedures of **module D: sensitivity analysis** is shown in Figure 7.1.

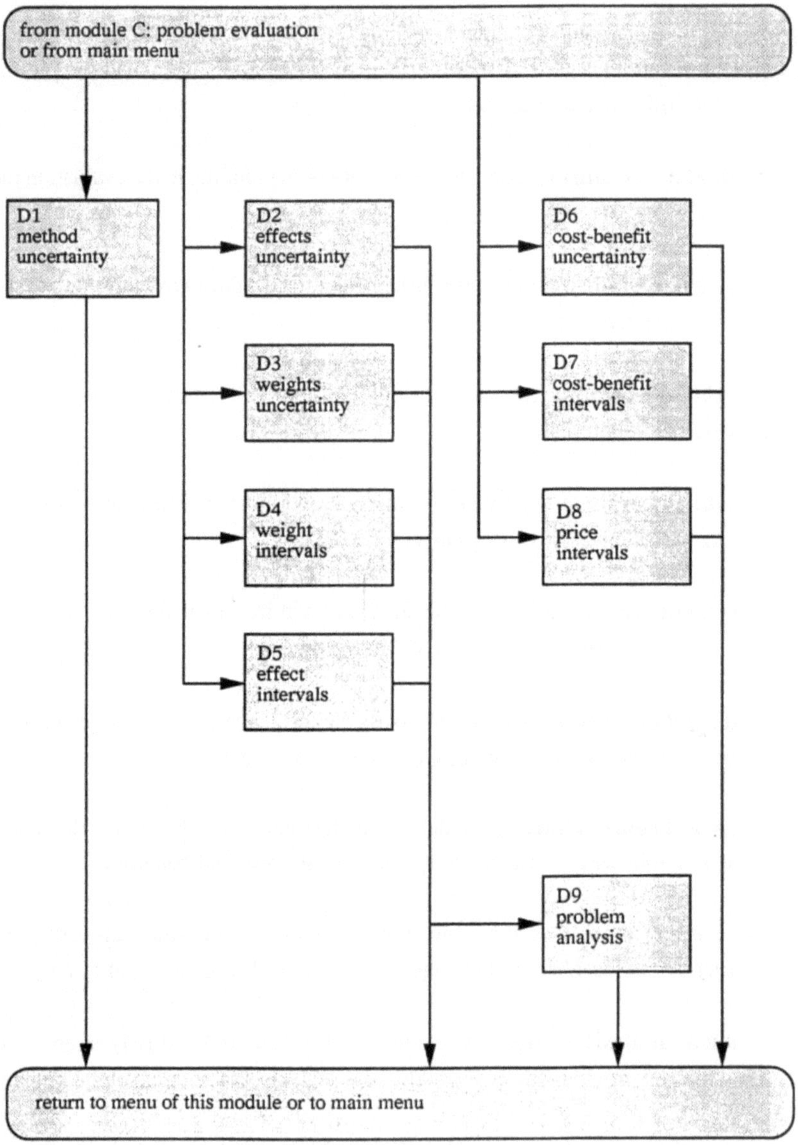

Figure 7.1. Procedures module D: sensitivity analysis.

DEFINITE menus

Upon entering the module, you will see the module menu. This menu provides access to the procedures available in this module. Each procedure is executed through its command line at the bottom of the screen.

There are two ways to make a selection from a menu.

1. Press **<SPACE>** to move through the menu and hit **<RETURN>** to select the highlighted option. An information line above the command line provides information on each highlighted option.

2. Hit the function key that corresponds to the option you want to select (e.g. **<F1>** to select the first option in a menu).

   In every command line you can:
   - select **Continue** to complete a procedure and to continue to the next step in the program;
   - hit **<ESC>** to cancel a procedure and return to the previous step;
   - use the **<CURSOR>** or **<ARROW>** keys to move the cursor within a table.

PROCEDURE D1. METHOD UNCERTAINTY

Purpose of this procedure

This procedure supports comparison of the results of problem evaluation.

Line graph

Your input and DEFINITE's output

Enter:

1. The **weights** linked to the results you want to analyze (if more than one set of weights is available);
2. **Line graph;**
3. The **alternatives** to compare (maximum 5).

This results in:

A line graph representing the results for the alternatives selected.

Bar graph

Your input and DEFINITE's output

Enter:

1. The **weights** linked to the results you want to analyze (if more than one set of weights is available);
2. **Bar graph.**

This results in:

A bar graph representing the results (select **rank** to rank the alternatives according to all results simultaneously; the method in the top row is considered the most reliable, etc.).

Example

Effects table. Improve transportation to Nyverhoeck.

effects	units	Highway	Road/bus	Train
Costs	mln $	-200	-250	-400
Travel time	min/prs/day	30	15	10
Capacity	mln/km/yr	20	30	40
Emissions NO_x	ton/yr	-1,000	-750	-100
Landscape	hectare	-10	-12	-2

Note that Travel time reflects the reduction in travel time (benefit) and Landscape reflects the loss of landscape (cost) that result from the alternatives.

Your input

Enter:

1. The **weights**. In this case the weights set: **wdirect.001** with information line: Standard. This is the weight set, Costs = 0.200, Travel time = 0.100, Capacity = 0.300, Emissions NO_x = 0.300, Landscape = 0.100;
2. **Line graph;**
3. **Alternatives.** In this case alternatives **Highway**, **Road/bus** and **Train**.

DEFINITE's output

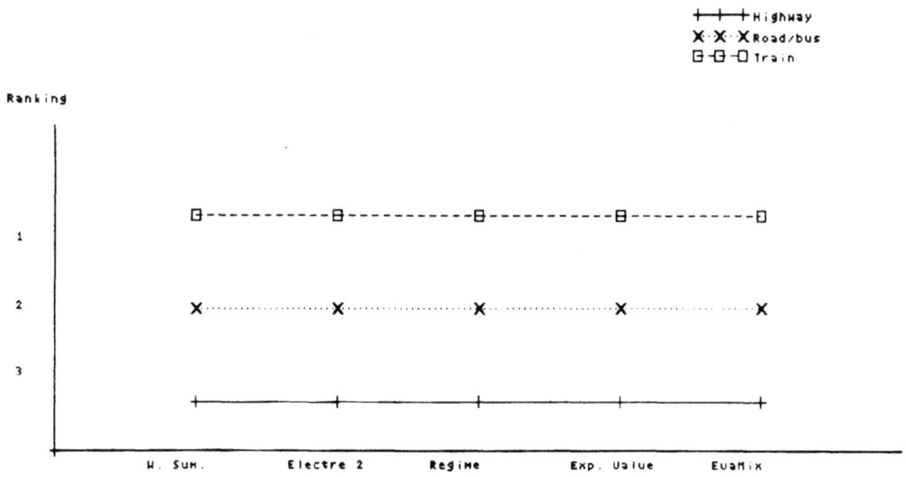

Figure 7.2. A line graph of the evaluation results.

For every method rank number 1 presents the best alternative, rank number 2 the second best, etc. If two alternatives rank equal, an intermediate rank number is assigned.

After selecting **conclusion**, DEFINITE presents a final ranking that is sufficiently certain. In this example:

Train > Road/bus > Highway

Comments

Selection of results

Only results that are calculated with the same weight set can be compared. Therefore you are asked to specify the weight set to which the results you want to analyze are linked. To do this it helps if you add a clear information line to your weights in module C. After you have selected a weight set DEFINITE searches for all results that match these weights. It is possible to save more than one result with the same weight set using Weighted Summation, for example. This can be useful if different types of standardization are used. If this is the case DEFINITE will ask which result needs to be included. An appropriate information line will help in selecting the required result. The Electre method may produce two or more partial rankings. If this is the case you are asked to select the partial ranking to include. An alternative marked with a ? will not be included.

Conclusion

To derive the conclusion of method uncertainty all alternatives are compared pair by pair. If in comparing two alternatives more than 70 percent of the methods applied rank one alternative over the other this alternative is assumed to rank over the other alternative with sufficient certainty. Note that in doing this, implicitly the assumption is made that all methods applied are equally suitable for the evaluation problem at hand.

The results of these comparisons are stored in a graph. A procedure of stepwise elimination is used to transform this graph into a ranking (see also the Electre 2 method).

Printing the result

Figures are *not* included in the report that can be generated in module E. It is therefore advisable to print this figure now. Procedures to save screen output are available for including figures in a text file. Users of WordPerfect 5.0 or higher, for example, can use the GRAB program to include this figure in a WordPerfect document. See the WordPerfect manual for further instructions.

PROCEDURE D2. EFFECTS UNCERTAINTY

Purpose of this procedure

This procedure assesses the sensitivity of a ranking obtained by one of the available multicriteria methods, for uncertainty in the scores of all effects.

Your input and DEFINITE's output

Enter:

1. A **result** of a multicriteria method to be analyzed;
2. An **uncertainty percentage** for each effect. This is the maximum percentage by which you expect the scores to deviate from the values included in the effects table. For example, if the score on an effect is 200, an uncertainty of 10 percent indicates that you are confident the score is between 180 and 220.

This results in:

A probability table. Each entry in this table represents the probability that an alternative receives a certain rank number, given the expressed uncertainty percentages on the effects scores.

Example

Effects table. Improve transportation to Nyverhoeck.

effects	units	Highway	Road/bus	Train
Costs	mln $	-200	-250	-400
Travel time	min/prs/day	30	15	10
Capacity	mln/km/yr	20	30	40
Emissions NO_x	ton/yr	-1,000	-750	-100
Landscape	hectare	-10	-12	-2

Note that Travel time reflects the reduction in travel time (benefit) and Landscape reflects the loss of landscape (cost) that result from the alternatives.

Your input

Enter:

1. The **result**. In this example the result **summ.001** with information line: direct assessment: capacity = 0,300, standard. This is the result obtained using Weighted

Summation and the weight set: Costs = 0.200, Travel time = 0.100, Capacity = 0.300, Emissions NO_x = 0.300, Landscape = 0.100;

2. **Uncertainty percentages**. In this example 25% for each of the effects.

DEFINITE's output

A probability table. Improve transportation to Nyverhoeck.

	Train	Road/bus	Highway
Rank 1	0.99	0.01	0.00
Rank 2	0.01	0.89	0.11
Rank 3	0.00	0.11	0.89
Total	2.98	1.90	1.11

This probability table shows that the ranking, given an uncertainty of 25%, is sufficiently reliable. The probability that Highway is the worst alternative (rank 3) equals 89%. The probability that Train is the best 99%.

After selecting **conclusion**, DEFINITE presents the ranking:
Train > Road/bus > Highway

Comments

Conclusion

The conclusion is based on the differences between the weighted sums of the probabilities of all rank numbers of alternatives that are adjacent in the ranking. In a problem with, for example, 10 alternatives rank 1 receives weight 10, rank 2 weight 9, etc.. The ranking of two alternatives is considered sufficiently certain if the difference between the weighted sums of these alternatives exceeds the arbitrarily set value of 0.2. The weighted sums are shown as total scores in the probability table.

Qualitative effects

This procedure can also be used if the effects table includes only qualitative or a combination of qualitative and quantitative effects. Instead of the percentage by which the quantitative score can deviate from the score in the effects table you are now asked the following question: "what is the probability that one randomly selected alternative should be one position higher in the ranking of alternatives for this effect?".

Calculation time

This procedure calculates the ranking of the alternatives a large number of times using a random generator. On slow computers this can cause some waiting time. If this procedure is used on results of methods that also use a random generator, such as the Regime method and the Random Values method or in combination with an appraisal table, waiting time can be considerable. The **<ESC>** key can be used to interrupt the calculations. Interrupting the calculations may lead to inaccurate results.

PROCEDURE D3. WEIGHTS UNCERTAINTY

Purpose of this procedure

This procedure assesses the sensitivity of a ranking obtained by one of the available multicriteria methods, for uncertainty in the scores of all weights.

Your input and DEFINITE's output

Enter:

1. A **result** of a multicriteria method to be analyzed;
2. An **uncertainty percentage** for each weight. This is the maximum percentage by which you expect the weights to deviate from the values included in the effects table. For example, if the weight assigned to an effect equals 0.5, an uncertainty of 10 percent indicates that you are confident the weight is between 0.45 and 0.55.

This results in:

A probability table. Each entry in this table represents the probability that an alternative receives a certain rank number, given the expressed uncertainty percentages on the weights.

Example

Effects table. Improve transportation to Nyverhoeck.

effects	units	Highway	Road/bus	Train
Costs	mln $	-200	-250	-400
Travel time	min/prs/day	30	15	10
Capacity	mln/km/yr	20	30	40
Emissions NO_x	ton/yr	-1,000	-750	-100
Landscape	hectare	-10	-12	-2

Note that Travel time reflects the reduction in travel time (benefit) and Landscape reflects the loss of landscape (cost) that result from the alternatives.

Your input

Enter:

1. The **result**. In this example the result **summ.001** with information line: direct assessment: capacity = 0,300, standard. This is the result obtained using Weighted Summation and the weight set: Costs = 0.200, Travel time = 0.100, Capacity = 0.300, Emissions NO_x = 0.300, Landscape = 0.100;
2. **Uncertainty percentages**. In this example 25% for each of the effects.

DEFINITE's output

Probability table. Improve transportation to Nyverhoeck.

	Train	Road/bus	Highway
Rank 1	1.00	0.00	0.00
Rank 2	0.00	1.00	0.00
Rank 3	0.00	0.00	1.00
Total	3.00	2.00	1.00

This probability table shows that the ranking, given an uncertainty of 25%, is reliable. The probability that Highway is the worst alternative (rank 3) equals 100%. The probability that Train is best is also 100%.

After selecting **conclusion**, DEFINITE presents the following ranking:
Train > Road/bus > Highway

Comments

Conclusion

The conclusion is based on the differences between the weighted sums of the probabilities to all rank numbers of alternatives that are adjacent in the ranking. In a problem with, for example, 10 alternatives rank 1 receives weight 10, rank 2 weight 9, etc.. The ranking of two alternatives is considered sufficiently certain if the difference between the weighted sums of these alternatives exceeds the arbitrarily set value of 0.2. The weighted sums are shown as total scores in the probability table.

Qualitative weights

This procedure can also be used if only qualitative weights are available. Instead of the percentage that the quantitative weight can deviate from the score in the effects table you are now asked the following question: "what is the probability that one randomly selected effect should be one position higher in priority order of effects?".

Calculation time

This procedure calculates the ranking of the alternatives a large number of times using a random generator. On slow computers this can cause some waiting time. If this procedure is used on results of methods that also use a random generator, such as the Regime method and the Random Values method, waiting time can be considerable. The **<ESC>** key can be used to interrupt the calculations. Interrupting the calculations may lead to inaccurate results.

PROCEDURE D4. WEIGHT INTERVALS

Purpose of this procedure

This procedure determines the intervals within which the rank order of two alternatives is insensitive to changes in weights. This procedure also determines which combination, most similar to the original combination of weights, reverses the order of two selected alternatives.

Weight interval

Your input and DEFINITE's output

Enter:

1. A **result** of a multicriteria method to be analyzed;
2. **Interval**;
3. Two **alternatives** to be compared;
4. The **weight** to be analyzed.

This results in:

The intervals of the selected weights.

Weight combination

Your input and DEFINITE's output

Enter:

1. A **result** of a multicriteria method to be analyzed;
2. **Combination**;
3. Two **alternatives** to be compared.

This results in:

A set of weights.

Example

Effects table. Improve transportation to Nyverhoeck.

effects	units	Highway	Road/bus	Train
Costs	millions	-200	-250	-400
Travel time	min/prs/day	30	15	10
Capacity	mln/km/yr	20	30	40
Emissions NO_x	ton/yr	-1000	-750	-100
Landscape	hectare	-10	-12	-2

Note that Travel time reflects the reduction in travel time (benefit) and Landscape reflects the loss of landscape (cost) that result from the alternatives.

Weight interval

Your input

Enter:

1. The **result** to be analyzed. In this example the result **exp.001** with information line: expected value: Capacity = 0.357. This is the result of the Expected Value method with weights: Emissions NO_x = 0.357, Costs = 0.157, Travel time = 0.065, Landscape = 0.065;
2. The **alternatives.** In this example **Highway** and **Train**;
3. One **weight.** In this example the weight of effect **Travel time**.

DEFINITE's output

Weight of Travel time

```
    1  - - - - -
       - - - - -
       - - - - -    Highway > Train
       - - - - -
       - - - - -
       - - - - -
       REVERSE      0.608
       + + + +
       + + + +
       + + + +
       + + + +
       + + + +
       + + + +
0.065  START        Train > Highway
       + + + +
    0  + + + +
```

If the weight of Travel time lies between 0 and 0.608, Train ranks over Highway. If the weight for Travel time is higher than 0.608 the ranking is reversed and Highway ranks over Train.

Since the sum of weights is always one, the weights of the other effects also change. The ratios between all other weights is kept constant. In this example this results in the following weights:

Reversal weight; ratios of all other weights are constant.

	Original	Reversal
Travel time	**0.065**	**0.608**
Capacity	0.357	0.150 [1]
Emissions NO_x	0.357	0.150
Costs	0.157	0.066
Landscape	0.065	0.027

If you select combination all weights will be allowed to change freely. In this case the weights most similar to all original weights that result in a reversal of the ranking of Train and Highway are determined.

Weight combination

Your input

Enter:

1. The **result** to be analyzed. In this example the result **exp.001** with information line: expected value: Capacity = 0.357. This is the result of the Expected Value method with weights: Emissions NO_x = 0.357, Costs = 0.157, Travel time = 0.065, Landscape = 0.065;
2. The **alternatives.** In this example **Highway** and **Train**;
3. **Combination.**

---

[1] The new weight of capacity $w_{capacity}$ can be calculated as follows:

$$w'_{capacity} = \frac{(1-w'_{traveltime})}{(1-w_{traveltime})} w_{capacity}$$

DEFINITE's output

The procedure checks all corner points (weight combinations where one weight is set equal to one) for a reversal in rank of the two selected alternatives. As a first intermediate result the procedure presents the corner points that result in rank reversal and the Euclidean distance between the original weights and these corner points.

The least distance corner point is selected. As a second intermediate result the reversal point on the line between the corner point and the original weights is presented. This is repeated for the corner point that has the next smallest distance.

In the next step the two reversal points that were found are used as a starting point to find the nearest reversal point (see also Section 4.4 of the book).

The final result is the weight combination that results in rank reversal of the two selected alternatives which is most similar to the original weights.

Weight combination (distance is: 0.327 [1]). Improve transportation to Nyverhoeck.

weights rank	score	alternatives rank	weights rank	score	alternatives rank
1. Capacity	0.357	1. Train	1. Costs	0.389	1. Road/bus
2. Emissions NO_x	0.357	2. Road/bus	2. Capacity	0.214	2. Highway
3. Costs	0.157	3. Highway	3. Emissions NO_x	0.186	Train
4. Travel time	0.065		4. Landscape	0.112	
5. Landscape	0.065		5. Travel time	0.100	

At these weights alternatives Highway and Train rank equal. Beyond this point Highway ranks over Train. Note that both alternatives now rank below Road/bus. The distance to the original weights = 0.327. It is clear that considerable changes in weights are required to reverse the ranking of these two alternatives.

[1] The distance d is defined as: $d = (\mathrm{Sum}(w'-w)^2)^{1/2}$.

PROCEDURE D5. EFFECT INTERVALS

Purpose of this procedure

This procedure determines the intervals within which the rank order of two alternatives is insensitive to changes in an effect score.

Your input and DEFINITE's output

Enter:

1. A **result** of a multicriteria method to be analyzed;
2. Two **alternatives** to be compared;
3. The **effect score** to be analyzed.

This results in:

The intervals of the selected effect score and alternatives.

Example

Effects table. Improve transportation to Nyverhoeck.

effects	units	Highway	Road/bus	Train
Costs	millions	-200	-250	-400
Travel time	min/prs/day	30	15	10
Capacity	mln/km/yr	20	30	40
Emissions NO_x	ton/yr	-1,000	-750	-100
Landscape	hectare	-10	-12	-2

Note that Travel time reflects the reduction in travel time (benefit) and Landscape reflects the loss of landscape (cost) that result from the alternatives.

Your input

Enter:

1. The **result**. In this example the result **regime.001** with information line : rank order: Capacity. This is the result of the Regime method and the weighting set: Capacity = Emissions NO_x, Costs, Travel time = Landscape;
2. The **alternatives** to be compared. In this example **Train** and **Highway** followed by **Train** and **Road/bus**;
3. The **score** to be analyzed. In this example **20** (Capacity of Highway) and **30** (Capacity of Road/bus).

DEFINITE's output

Capacity of Highway

```
MAX     - - - - - -
        - - - - - -
        - - - - - -    Highway > Train
        - - - - - -
        - - - - - -
        - - - - - -
        REVERSE        43.44
        + + + + +
        + + + + +
        + + + + +
20.00   START          Train > Highway
        + + + + +
        + + + + +
        + + + + +
        + + + + +
0       + + + + +
```

Capacity of Road/bus

```
MAX     - - - - - -
        - - - - - -
        - - - - - -    Road/bus > Train
        - - - - - -
        - - - - - -
        - - - - - -
        - - - - - -
        REVERSE        40.88
        + + + + +
30.00   START          Train > Road/bus
        + + + + +
        + + + + +
        + + + + +
        + + + + +
        + + + + +
0       + + + + +
```

The first interval shows that if the Capacity of Highway is between 0 and 43.44 alternative Train is preferred to alternative Highway, and that if Capacity is higher than 43.44 Highway is preferred to Train.

The second interval shows that if the Capacity of Road/bus is between 0 and 40.88 alternative Train is preferred to alternative Road/bus, and that if Capacity is higher than 40.88 Road/bus is preferred to Train.

PROCEDURE D6. COST-BENEFIT UNCERTAINTY

Purpose of this procedure

This procedure assesses the sensitivity of a ranking, obtained with Cost-Benefit analysis for uncertainties in effect scores.

Your input and DEFINITE's output

Enter:

1. **Benefit-cost ratio** or **net present value**;
2. An **uncertainty percentage** for each effect. This is the maximum percentage by which you expect the scores to deviate from the values included in the effects table. For example, if a score is 200, an uncertainty of 10 percent indicates that you are confident the score is between 180 and 220.

This results in:

A probability table. Each entry in this table represents the probability that an alternative obtains a certain rank number, given the expressed uncertainty percentages on the effects scores.

Example

First the results of Cost-Benefit analysis are presented.

Results of Cost-Benefit analysis (scores in millions). Improve transportation to Nyverhoeck.

alternative discount rate: 5.00%	net present value	benefits/ costs	internal rate of return (%)
1. Highway	-89.09	0.72	negative
2. Road/bus	-278.18	0.29	negative
3. Train	-380.00	0.19	unknown

Your input

Enter:

1. **Net present value**

Next the cost-benefit table is presented and you are asked to specify an uncertainty percentage for each effect.

Cost-benefit table. Improve transportation to Nyverhoeck.

effects	Cost Benefit	units	prices millions	Highway	Road/bus	Train
Costs	C	millions	1	218.18	272.73	436.36
Landscape	C	hectare	10	10.00	12.00	2.00
Emissions NO_x	C	ton/yr	?	1909.09	1431.82	190.91
Travel time	B	min/prs/day	4	57.27	28.64	19.09
Capacity	B	mln/km/yr	?	38.18	57.27	76.36

Note that Travel time reflects the reduction in travel time (benefit) and Landscape reflects the loss of landscape (cost) that result from the alternatives.

Enter:

2. The **uncertainty percentages**. In this example **25%** for each effect.

DEFINITE's output

A probability table. Improve transportation to Nyverhoeck.

	Train	Road/bus	Highway
Rank 1	1.00	0.00	0.00
Rank 2	0.00	0.97	0.03
Rank 3	0.00	0.03	0.97
Total	3.00	1.97	1.03

After selecting **conclusion**, DEFINITE presents the following ranking:

Train > Road/bus > Highway

This probability table shows that the ranking, given an uncertainty of 25%, is reliable. The probability that Highway is the best alternative (rank 1) equals 100%. The probability that Train is the worst alternative equals 97%.

Comments

Conclusion

The conclusion is based on the differences between the weighted sums of the probabilities of all rank numbers of alternatives that are adjacent in the ranking. In a problem with, for example, 10 alternatives rank 1 receives weight 10, rank 2 weight 9, etc.. The ranking of two alternatives is considered sufficiently certain if the difference between the weighted sums of these alternatives exceeds the arbitrarily set value of 0.2. The weighted sums are shown as total scores in the probability table.

Calculation time

This procedure calculates the ranking of the alternatives a large number of times using a random generator. On slow computers this can cause some waiting time. The **<ESC>** key can be used to interrupt the calculations. Interrupting the calculations may lead to inaccurate results.

PROCEDURE D7. COST AND BENEFIT INTERVALS

Purpose of this procedure

This procedure determines the intervals within which the rank order of two alternatives is insensitive to changes in an effect score.

Your input and DEFINITE's output

Enter:

1. **Benefit-cost ratio** or **net present value;**
2. Two **alternatives** to be compared;
3. The **effect score** to be analyzed.

This results in:

The intervals of the selected effect score.

Example

First the results of Cost-Benefit analysis are presented.

Results of Cost-Benefit analysis (scores in millions). Improve transportation to Nyverhoeck.

alternative discount rate: 5.00%	net present value	benefits/ costs	internal rate of return (%)
1. Highway	-89.09	0.72	negative
2. Road/bus	-278.18	0.29	negative
3. Train	-380.00	0.19	unknown

Your input

Enter:

1. **Net present value;**

Next the discounted effects table is presented and you are asked to specify the alternatives to be compared and the score to be analyzed.

Discounted effects table. Improve transportation to Nyverhoeck.

effects	Highway	Road/bus	Train
Costs	218.18	272.73	436.36
Landscape	10.00	12.00	2.00
Emissions NO_x	1909.09	1431.82	190.91
Travel time	57.27	28.64	19.09
Capacity	38.18	57.27	76.36

Note that Travel time reflects the reduction in travel time (benefit) and Landscape reflects the loss of landscape (cost) that result from the alternatives.

Enter:

2. The **alternatives**. In this example **Highway** and **Train**;
3. The **cost or benefit score**. In this example the loss of **Landscape** caused by alternative Highway.

DEFINITE's output

Landscape of Highway

```
  0   + + + + +
      + + + + +
-10   START      Highway > Train
      + + + + +
      + + + + +
      + + + + +
      + + + + +
      REVERSE    -39.06
      - - - - - -
      - - - - - -     Train > Highway
      - - - - - -
MIN   - - - - - -
```

This cost interval shows that if the loss of landscape caused by alternative Highway is between 0 and 39.06 hectares alternative Highway is preferred to alternative Train, and that if the loss of landscape is higher than 39.06 hectare Train is preferred to Highway.

PROCEDURE D8. PRICE INTERVALS

Purpose of this procedure

This procedure determines the intervals within which the rank order of two alternatives is insensitive to changes in a price.

Your input and DEFINITE's output

Enter:

1. **Benefit-cost ratio** or **net present value;**
2. Two **alternatives** to be compared;
3. The **price** to be analyzed.

This results in:

The intervals of the selected price.

Example

First the results of Cost-Benefit analysis are presented.

Results of Cost-Benefit analysis (scores in millions). Improve transportation to Nyverhoeck.

alternative discount rate: 5.00%	net present value	benefits/ costs	internal rate of return (%)
1. Highway	-89.09	0.72	negative
2. Road/bus	-278.18	0.29	negative
3. Train	-380.00	0.19	unknown

Your input

Enter:

1. **Net present value;**

Next the cost-benefit table is presented and you are asked to specify the alternatives to be compared and the price to be analyzed.

Cost-benefit table. Improve transportation to Nyverhoeck.

effects	Cost Benefit	units	prices millions	Highway	Road/bus	Train
Costs	C	millions	1	218.18	272.73	436.36
Landscape	C	hectare	10	10.00	12.00	2.00
Emissions NO_x	C	ton/yr	?	1909.09	1431.82	190.91
Travel time	B	min/prs/day	4	57.27	28.64	19.09
Capacity	B	mln/km/yr	?	38.18	57.27	76.36

Note that Travel time reflects the reduction in travel time (benefit) and Landscape reflects the loss of landscape (cost) that result from the alternatives.

Enter:

2. The **alternatives.** In this example **Highway** and **Train**;
3. The **price**. In this example **10** (the price of one hectare of Landscape).

DEFINITE's output

Price of Landscape

MAX	- - - - -	
	- - - - -	
	- - - - -	Train > Highway
	- - - - -	
	- - - - -	
	REVERSE	46.56
	+ + + + +	
	+ + + + +	
	+ + + + +	
10.00	START	Highway > Train
	+ + + + +	
	+ + + + +	
	+ + + + +	
0	+ + + + +	

This price interval shows that if the price assigned to a hectare loss of landscape is between 0 and 46.56 million, alternative Highway is preferred to alternative Train, and that at any higher price Train is preferred to Highway.

PROCEDURE D9. PROBLEM ANALYSIS

Purpose of this procedure

This procedure combines the conclusions of various procedures of sensitivity analysis that are performed on a result of a multicriteria method. The procedure checks whether for a selected multicriteria result all relevant procedures have been applied. The procedure determines the ranking that is sufficiently certain according to both weights and effects uncertainty.

Your input and DEFINITE's output

Enter:

The **result** of problem evaluation to be analyzed.

This results in:

Advice on additional analyses to be carried out.

or

An assessment of the reliability of the selected result.

Example

Effects table. Improve transportation to Nyverhoeck.

effects	units	Highway	Road/bus	Train
Costs	millions	-200	-250	-400
Travel time	min/prs/day	30	15	10
Capacity	mln/km/yr	20	30	40
Emissions NO_x	ton/yr	-1,000	-750	-100
Landscape	hectare	-10	-12	-2

Note that Travel time reflects the reduction in travel time (benefit) and Landscape reflects the loss of landscape (cost) that result from the alternatives.

Your input

Enter:

The **result**. In this example the result **summ.001** with information line: direct assessment: capacity = 0,300, standard. This is the result obtained using Weighted Summation and the weight set: Costs = 0.200, Travel time = 0.100, Capacity = 0.300, Emissions NO_x = 0.300, Landscape = 0.100.

DEFINITE's output

If you have not applied and saved the results of weights uncertainty and effects uncertainty you are instructed to do so.

In this example these results were saved. First the results of weights uncertainty and effects uncertainty are presented. In both cases the ranking Train > Road/bus > Highway proved to be sufficiently certain.

The result of problem analysis is therefore that this ranking is sufficiently certain.

8. REPORT GENERATION

Introduction

After all relevant analyses are performed, this module can be used to produce an evaluation report.

Report generation is supported by the following procedures:

E1. **Generate the report**. This procedures combines all results saved within a session into a standard evaluation report.

E2. **Edit the report** can be used to edit the standard evaluation report using the build in word processor MicroStar.

A flow diagram of the procedures of **module E: report generation** is shown in Figure 8.1.

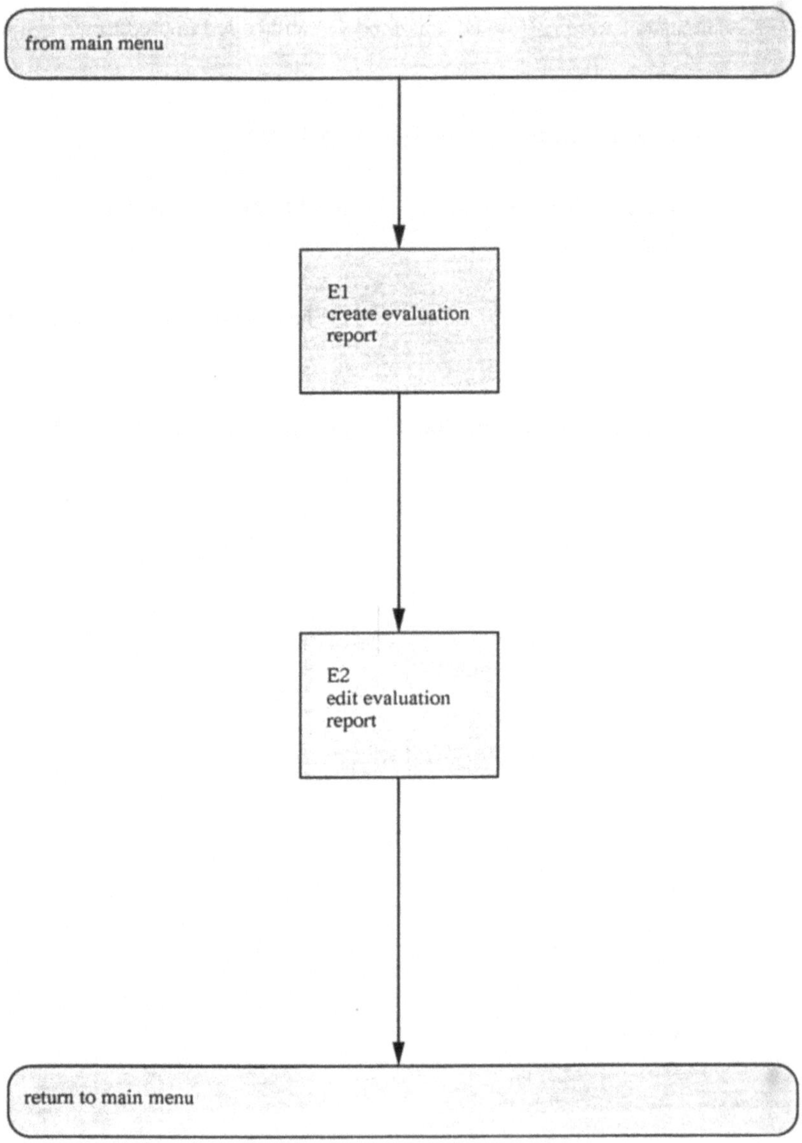

Figure 8.1. Procedures module E: report generation.

PROCEDURE E1. GENERATE THE REPORT

Purpose of this procedure

This procedure combines all results saved within a session into a standard evaluation report.

Description

DEFINITE allows you to save your results in text files in two ways:

1. The modules can be saved in different files. These files can be edited immediately using the built in MicroStar editor.
2. The modules can be saved together in a single file. This file can be edited with your own editor. The file has to be loaded as a DOS-text or ASCII FILE.

Your input and DEFINITE's output

Enter:

How you want your text file to be saved:

1. For editing within DEFINITE;
2. For editing outside DEFINITE.

This results in:

A standard evaluation report.

Comments

Importing an ASCII file.

To import an ASCII file generated with option 2 into a word processor you need to use the correct import procedure. In WordPerfect, for example, this procedure is as follows:

1. Select Text in/out:	**<CTRL><F5>**
2. Select **Dos text**:	**1**
3. Select **Load with Srt in H-zone**:	**3**

PROCEDURE E2. EDIT THE REPORT

Purpose

This procedure can be used to edit the standard evaluation report.

Description

The five chapters of the standard evaluation report are saved in five separate files. You have now entered the DEFINITE editing menu. From this menu you have various options.

Enter:

1 to edit Chapter 1: Introduction
2 to edit Chapter 2: Problem definition
3 to edit Chapter 3: Problem presentation
4 to edit Chapter 4: Problem evaluation
5 to edit Chapter 5: Sensitivity analysis
6 to print a chapter
7 to copy a chapter to a different directory or diskette
8 to quit MicroStar and return to the main menu

This menu includes the minimum options you need to edit and print your report. After you have selected one of the chapters you can reactivate this menu with **<ALT>9**.

When you enter a new chapter or when you leave MicroStar you are asked to save the last chapter you have edited. You can use **<F2>** to save a chapter at any other moment. This is, however, not essential.

With **<F10>** you enter the MicroStar menu. The DEFINITE **<ALT>** options are not available from the MicroStar menu. Hit **<ESC>** to return to your text. Select **M H** (MicroStar help) for help on the various MicroStar options. Always exit MicroStar through the DEFINITE editing menu.

Part 3

Examples from the practice of environmental decision making

9. INTRODUCTION TO THE EXAMPLES

This part of the DEFINITE book is included to demonstrate the potential of DEFINITE, using examples from the practice of environmental decision making. After installation the examples are available on your hard disk for direct use within the program. The instructions in the following four chapters, one chapter for each of the examples, will lead you step by step through them. The best way to get to know DEFINITE is first to go through the examples step by step using the descriptions in the next four chapters. Next you could experiment with the examples, and finally enter your own example. The example sessions are protected. To be able to use an example you have to create your own session and copy the example to your session. It is therefore impossible to do permanent damage to any of the examples.

The four examples are:

1. **Forest**
   The Forest and Timber Inquiry in Australia.
   In this example five alternatives for forest and timber use in Australia are compared according to six criteria. Multicriteria analysis combined with diagrams are used to compare these alternatives.

2. **Soil**
   Soil conservation in Maphutseng, Lesotho.
   In this example Cost-Benefit analysis is used to determine whether a soil conservation project is worthwhile. In other words the project alternative is evaluated against the no project alternative.

3. **Energy**
   Environmental effects of electricity production in the Netherlands.
   The effects table in this example includes a large number of effects. These effects are measured on a quantitative scale but cannot be valued in monetary terms. A combination of graphic presentation, aggregation and multicriteria analysis is used to evaluate the alternatives.

4. **Nuclear**
   Siting nuclear plants in the Netherlands.
   Many locations need to be evaluated on many criteria in this example. All but one of the evaluation criteria in this example are measured on a qualitative measurement scale. A combination of multicriteria analysis and sensitivity analysis is used to select the best location.

The descriptions of the four examples include:

- a description of the example;
- screen outputs;
- explanations of the screen outputs and most importantly;
- "what to do: boxes".

The "what to do: boxes" will tell you which keys to hit to go through the examples in the way intended in the descriptions. All other keys are of course available for experimentation but we suggest that in the first round you stick to the suggestions given. The box below includes, as an example, the instructions for starting the Forest example.

What to do:

Go to the DEFINITE directory:	**CD DEFINITE**	**<RETURN>**
Start the program.	**DEFINITE**	**<RETURN>**
Enter your own session name:	**YOURNAME**	**<RETURN>**
Answer yes to the question:	**Y**	
Select the example **Forest**:	**<CURSOR><CURSOR>**	**<RETURN>**

10. THE FOREST AND TIMBER INQUIRY IN AUSTRALIA

10.1. Introduction

This example is based on an evaluation study conducted by the Resource Assessment Commission of Australia to support the Forest and Timber Inquiry.[1]

The Inquiry's terms of reference require it to identify and evaluate options for the use of Australia's forest and timber resources. Australia now has 43 million hectares of forested land, covering 5 percent of the total land area. This equals 65 percent of the area of forested land that existed at the time of European settlement. Another 92 million hectares of woodland cover a further 14 percent of the country (see Figure 10.1). Regrowth from past logging covers about 60 percent of the forest area. The remaining 40 percent is either not logged or has been lightly logged. About 6 million hectares of this relatively undisturbed area is in designated nature reserves.

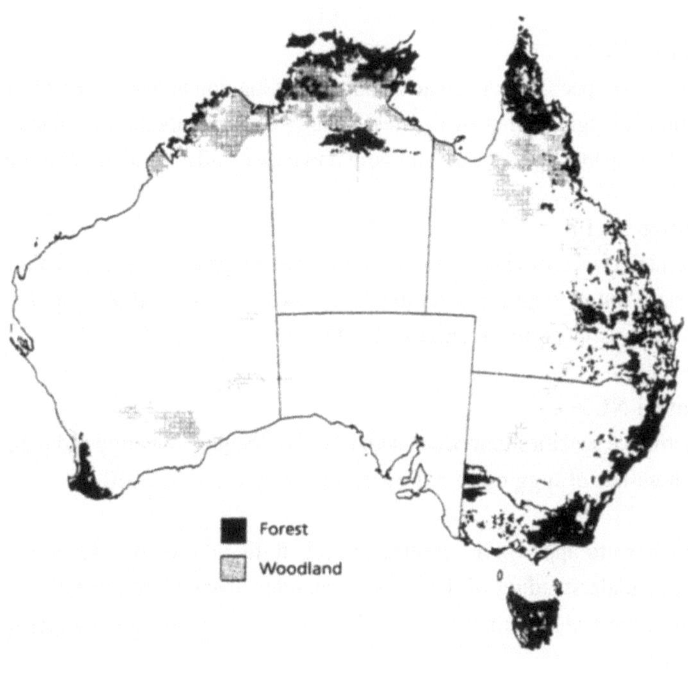

Figure 10.1. Forest and woodland in Australia, 1990 (Source: RAC, 1992).

---

[1] Resource Assessment Commission (1992). *Forest and Timber Inquiry*, Final Report. Australian Government Publishing Service, Canberra.

The Inquiry specified six alternatives and used a computer model to predict the effects of these alternatives.

1. **Alternative CT**

   The 'current trends' alternative which adopts existing assumptions for future forest and timber resource use as described by the forest management agencies. It provides the baseline for comparison of the other alternatives.

2. **Alternative HII**

   The 'high industry investment' alternative. This alternative allows for the establishment of a significant, intensively managed hardwood and softwood plantation resource, and for intensive management of suitable native forest types. Investment in pulp and paper processing facilities is greater than that projected for the 'current trends' alternative.

3. **Alternative MII**

   The 'moderate industry investment' alternative. This alternative does not allow for intensive management practices in native forest, and pulp and paper processing facilities lie between those proposed for the 'current trends' alternative and those proposed for alternative HII.

4. **Alternative NLM**

   This alternative specifies immediate withdrawal of logging in mature and old-growth forest but continued logging in regrowth forests. Hardwood plantations are expanded, the export of hardwood woodchips ceases, and pulp and paper mills are of small to medium scale.

5. **Alternative NL10**

   This alternative specifies immediate withdrawal of logging in mature and old-growth forest and a phase-out of logging in regrowth forests by the year 2000. Other specifications are the same as those for alternative NLM.

6. **Alternative NL30**

   This alternative specifies immediate withdrawal of logging in mature and old-growth forest and a phase-out of logging in regrowth forests by the year 2020.

The Inquiry chose to apply multi-criteria analysis to the alternatives for two reasons:

- to improve understanding of the relative qualities of the alternatives;
- to evaluate the technique's usefulness for comparing and evaluating alternative uses of natural resources.

The Resource Assessment Commission acquired DEFINITE in order to evaluate the usefulness of multicriteria methods for evaluating alternative uses of Australia's forests.

In this application of multicriteria analysis reference years (2000 and 2030) are used to compare the expected state of the forest resource and the associated industry in future years. This approach is in accord with the concept of sustainable development which aims to maintain both economic and environmental quality indefinitely. Multicriteria analysis, however, contrasts with most applications of Cost-Benefit analysis in that, in the latter case, expected benefits and costs resulting from an alternative in all future years are discounted to their net present value.

This application includes four basic steps.

1. **Problem definition**
   In this step the effects table is created. The effects table is made up of the selected set of alternatives as columns, and a number of criteria that reflect the predicted quality of the alternatives in specific years in the future, as rows.

2. **Problem presentation**
   This step involves aggregating the criterion scores from the effects table into two categories: an index of economic quality and an index of environmental quality. This presentation makes explicit the trade-off between economic and environmental qualities.

3. **Problem evaluation**
   The third step requires the use of a multicriteria method to rank the alternatives.

4. **Sensitivity analysis**
   This step involves applying sensitivity analysis to determine the effect of changing weights and scores on the evaluation results.

What to do:

Go to the DEFINITE directory:	**CD DEFINITE**	**<RETURN>**
Start the program:	**DEFINITE**	**<RETURN>**
Enter your own session name:	**YOURNAME**	**<RETURN>**
Answer yes to the question:	**Y**	
Select the example **Forest**:	**<CURSOR><CURSOR>**	**<RETURN>**

It may take some time to go through the whole example. Return to the main menu and exit the program if you want to stop before reaching the end of the example. When you start the example again you can continue where you stopped.

You have now entered the main menu.

10.2. Problem definition

Problem definition is the first and possibly the most important step in multicriteria analysis. It involves identification of the feasible alternatives, identification of the evaluation criteria, and assessment of criterion scores. Alternatives, criteria and criterion scores are included in the effects table. The path this example takes through module A is emphasized in Figure 10.2.

Criteria

The following three criteria were selected to reflect the economic quality of the alternatives:

1. **Value added** - this criterion is defined as the sum of wages and profits in the forest and forest products industry. It is measured in millions of dollars and reflects the contribution of the forest and forest products industry to gross domestic product.
2. **Direct employment** is defined as the number of persons directly employed in the forest and forest products industry and reflects the job creation prospects in the economy.
3. **Balance of trade** - this criterion is defined as forest product exports minus forest product imports. It is measured in millions of dollars.

The following three criteria represent the environmental quality of the alternatives:

4. **Old-growth forest unused** - this criterion is expressed in hectares.
5. **Unavailable for logging** - this criterion is expressed in hectares and reflects the sum of the areas of each forest group that have been allocated to conservation reserves and the area of each forest group that is unavailable for logging in production forest tenures.
6. **Soil and water quality** - this criterion is expressed as a qualitative index (plus or minus) and reflects an assessment of the predicted impact of human disturbance on soil and water quality under each of the six alternatives. It is derived subjectively, by expert judgement.

Projections were made for a large number of response variables or indicators, such as log yield, intermediate and final products, and production areas, for each decade from 1990 to 2090. For the purpose of the multicriteria analysis the year 2000 is selected to represent the expected short-term qualities of the alternatives, and the year 2030 is chosen to represent the medium term. The lack of reliability of projections of consumption trends does not permit analysis of the long-term qualities of the available alternatives. In this example only the results for the year 2030 are presented.

What to do:

Enter module A: **problem definition:**	**<F1>**
Enter procedure A1: **create an effects table:**	**<F1>**

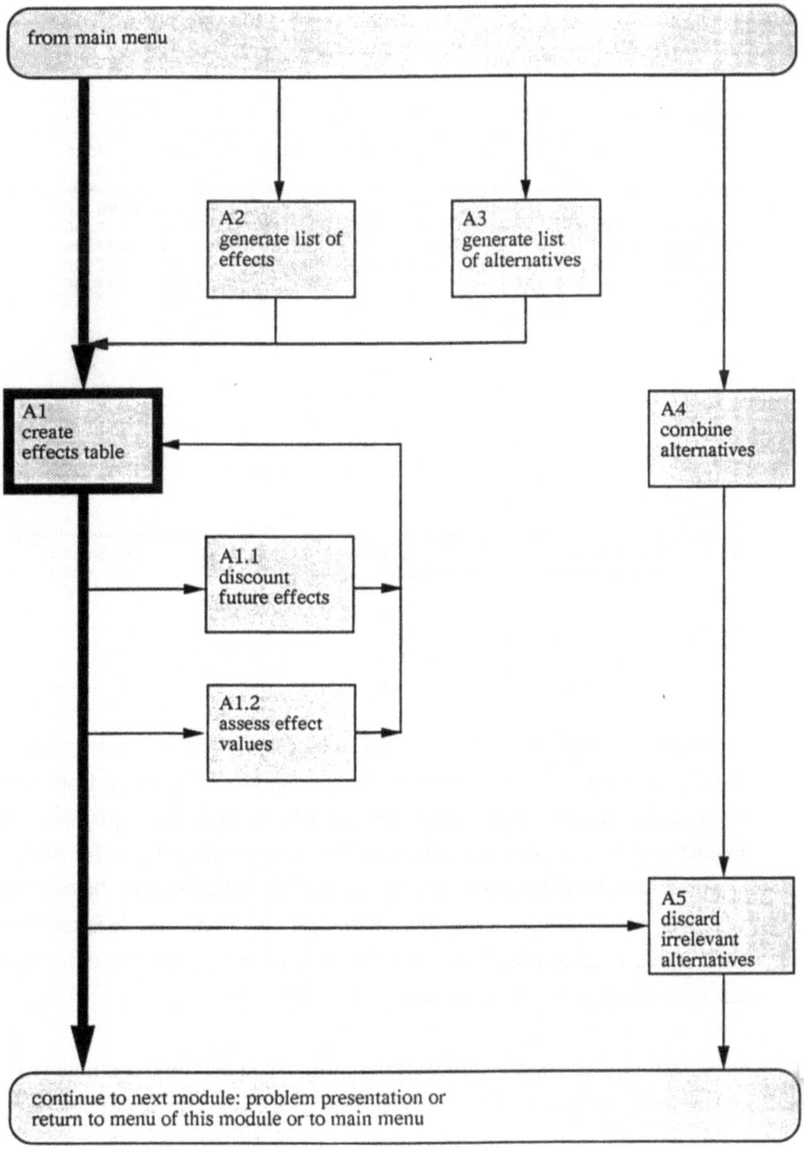

Figure 10.2. Problem definition. Forest and Timber Inquiry Australia.

Information screens are presented upon entering each procedure. Hit **<ANY KEY>** to continue. The instruction to hit any key is not included in the "what to do: boxes".

This results in the following screen.

Screen 10.1. Effects table. Forest and Timber Inquiry Australia.

effects table	CT	HII	MII	NLM	NL10	NL30
Year	2030	2030	2030	2030	2030	2030
Value added	4819.00	5688.00	5524.00	4733.00	4473.00	4490.00
DirectEmployment	382.00	448.00	442.00	377.00	366.00	367.00
Balance of trade	-1427.40		-320.72	-1316.60	-1493.00	-1447.20
OldGrowth unused	2728.00	2728.00	2728.00	3062.00	3062.00	3062.00
Unav.for logging	2347.10	2347.10	2347.10	2389.30	3444.30	3444.30
Soil/water qual.	+	--	+	+	++	++

```
Select an option:
=Add an alternative=====================================================A1  ==
F1.alternative  F2.effect      F3.descriptor   F4.edit        F5.units
F6.time         F7.information F8.help         F9.continue    F10.more
```

The tables illustrate two noteworthy points. First, multicriteria analysis permits the use of criteria having different units of measurement. In this analysis units range from measured variables such as dollar values, employment statistics and forest areas to qualitative indices that are expressed as plus or minus signs. Second, the absolute differences for many of the criteria are small across the alternatives and for some criteria do not vary between subsets of alternatives. This is a consequence of the alternative specifications. It is clear that the criterion values represented here will also be influenced to some extent by other sectors of the Australian economy and by the activities of other countries.

The effects table is the final result of problem definition. You can now continue to the next module: problem presentation.

What to do:

Select **continue** to leave this procedure:	**<F9>**
Continue to module B: **problem presentation**:	**<F2>**

10.3. Problem presentation

The purpose of problem presentation is to present the effects table in a form that supports comparison of the alternatives without applying a formal decision rule. The path this example takes through module B is emphasized in Figure 10.3.

Graphic presentation

What to do:

Select **graphic presentation**:	<F3>

This results in the following screen.

Screen 10.2. Graphic presentation of the effects table.

Forest and Timber Inquiry Australia.

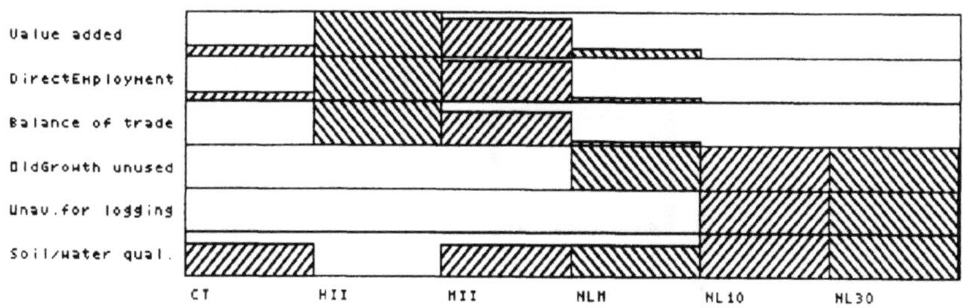

Select an option:
=Exchange two rows =
F1.exchange row F2.exchange clm F3.print F4.rank F5.information
F6.Help F7.continue

To produce Screen 10.2 the scores are standardized between zero (the worst alternative), and one (the best alternative) for each category. The highest bar represents the best score.

The alternatives can be ranked graphically by exchanging rows and columns. In Screen 10.3 the rows are ordered according to the priority assigned to each category, and the columns

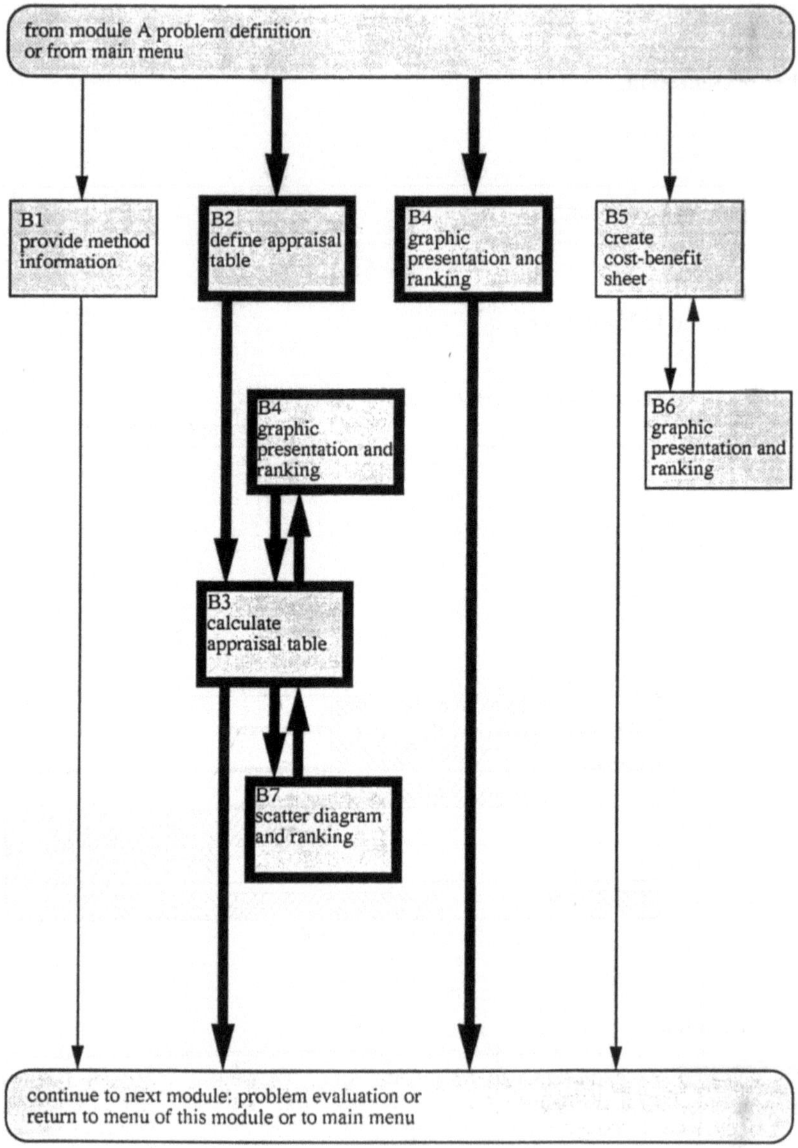

Figure 10.3. Problem presentation. Forest and Timber Inquiry Australia.

automatically using a combination of the Expected Value method and Weighted Summation. To produce the next screen priority is given to environmental quality.

What to do:

Exchange rows to move the environmental criteria to the top: **<F1>**
Rank the alternatives: **<F4>**

Screen 10.3. Graphical ranking of the effects table, priority given to environmental criteria. Forest and Timber Inquiry Australia.

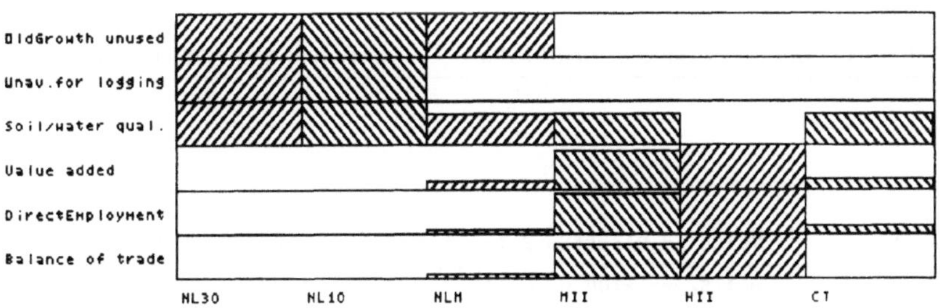

Select an option:
=Rank the alternatives
F1.exchange row F2.exchange clm F3.print F4.rank **F5.information**
F6.Help F7.continue

Screen 10.3 shows the clear trade-off between economic and environmental quality for the alternatives examined: a high economic quality is paid for by a low environmental quality and vice versa. Screen 10.3 also shows that changing the order of the criteria within either the environmental quality group or the economic quality group will have little influence on the ranking: exchanging, for example, the added value and balance of trade criteria will not alter the ranking of the alternatives.

Indices for economic and environmental quality

The second step in the multicriteria analysis is to aggregate the criteria into two index categories. In this section the three economic criteria - Value added, Direct employment, and Balance of trade - are combined into a single index of economic quality. In the same way the three environmental criteria - Old-growth forest unused, area Unavailable for logging, and Soil and water quality - are combined into a single index of environmental quality. With these indices, the trade-off between environmental and economic aspects of the alternatives can be examined.

What to do:

Continue :	**<F7>**
Select **appraisal table**:	**<F2>**
Select **calculate**:	**<F1>**

Screen 10.4. Definition of the appraisal table. Forest and Timber Inquiry Australia.

definitions	Value ad	DirectEm	Balance	OldGrowt	Unav.for	Soil/wat
units	mill$/yr	x100 p.	mill$/yr	x1000 ha	x10000ha	---/+++
Economic quality	0.33	0.33	0.33			
Environ. quality				0.33	0.33	0.33

```
Select an option:
═Calculate the appraisal table══════════════════════════════B3 ══
F1.calculate    F2.edit
```

Screen 10.5. Appraisal table. Forest and Timber Inquiry Australia.

appraisal	CT	HII	MII	NLM	NL10	NL30
Economic quality	0.17	1.00	0.86	0.16		0.02
Environ. quality	0.26	0.14	0.26	0.61	1.00	1.00

```
Select an option:
==Edit the definitions table=======================================B3  ==
F1.definitions  F2.move          F3.graph       F4.diagram     F5.units
F6.print        F7.information   F8.help        F9.continue
```

The criterion scores are transformed into an economic quality index and an environmental quality index in two steps. First, all scores are standardized between zero and 100. Next, the indices are calculated as the weighted sums of these standardized scores. The weights represent the relative importance of the criteria within the index. This is a crucial part of multicriteria analysis because the weightings assigned to criteria can have a significant impact on the final alternative rankings. For this reason it is usually desirable to obtain assessments of the weights assigned to criteria using either expert opinion, or community interest group opinion.

As shown in relation to Screen 10.3, and as shown using sensitivity analysis, in this analysis the sensitivity of the ranking of the alternatives to changes in the criterion weights is limited. In this application of the multicriteria analysis technique it was therefore not necessary to obtain a precise assessment of the criterion weights, and the economic index and environmental index for each criterion were therefore given equal weight.

There are two possibilities to calculate the appraisal scores from the effects:

1. The appraisal scores are calculated as the weighted sums of the effects. This is the default setting.
2. The appraisal scores are calculated as the weighted sums of the standardized effects. This is the setting used in this example.

To combine scores measured in different measurement units or measurement scales, these scores need to be standardized before they can be aggregated. In this example the effects are standardized. To change the setting of the appraisal table you need to go to utilities from the main menu. The utilities screen will be shown at the end of this example.

What to do:

Select **graph**:	**<F3>**

Screen 10.6. Graphical presentation of the appraisal table.

Forest and Timber Inquiry Australia.

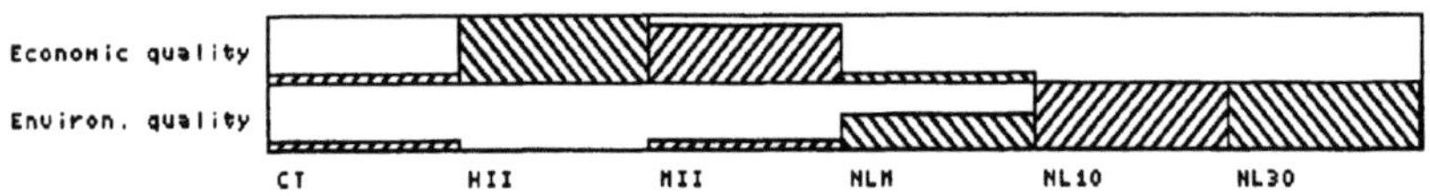

Select an option:
=Exchange two rows
F1.exchange row F2.exchange clm F3.print F4.rank F5.information
F6.Help F7.continue

The trade-off between economic and environmental quality

In this section the values of the two indices are presented in a diagram. This diagram is used to explore the trade-off between economic and environmental quality.

What to do:

Leave the graph:	**<F7>**
Select **diagram**:	**<F4>**
Select the two indicators:	**<F1><F2>**
Select **numbers**:	**<F1>**
Select **line**:	**<F5>**
Move line:	**<F6>**

This results in the following screen:

Screen 10.7. Graphical ranking of the alternatives, equal weight, year 2030.
Forest and Timber Inquiry Australia.

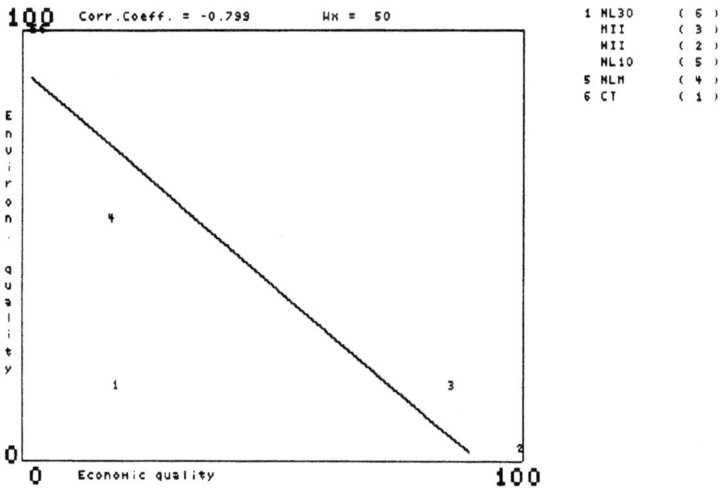

Move line
Move line with cursor keys.
Return to menu with <ESC>

The ranking of the alternatives is shown at the right hand side of the diagram. The first position is shared by four alternatives: NL30, MII, HII and NL10. NLM ranks fifth and CT sixth. The numbers in brackets behind each alternative refer to the numbers in the diagram. For example, the 1 at the lower left corner of the diagram refers to CT. Numbers 6 and 5 (NL30 and NL10) are in the upper left corner but, because they have an environmental score of 100, they coincide with the border of the diagram and are therefore difficult to see.

Screen 10.7 shows the economic and environmental quality of the alternatives in the year 2030, as calculated in the appraisal table. The horizontal axis represents economic quality; the vertical axis represents environmental quality. The ideal option would combine maximum economic quality with maximum environmental quality. This ideal option would have scored 100 for both economic and environmental quality and would therefore be found in the top right-hand corner of the diagram. The alternatives are ranked on the bases of the distance from the ideal option. This multicriteria method is known as the 'Ideal Point' method (Janssen, 1992).

Various measures are available to determine the distance of the alternatives from the ideal point. In this application, the distance is measured as the sum of the distances along the horizontal and vertical axes. The diagonal in Screen 10.7 intersects the economic and environmental quality axes at equal distance from the ideal point. The slope of the line indicates that equal weight is given to economic and environmental quality. Since all points on the line share the same distance from the ideal point they all have the same position in the ranking.

The alternatives can now be ranked visually by moving the line from top right to bottom left. The first alternative to cross the line, and therefore the most preferred alternative, is alternative NL30, immediately followed by alternatives MII and HII.

The correlation coefficient listed at the top of the diagram indicates the extent to which it is necessary to trade off economic and environmental quality. A value close to one indicates a minimal need to trade off both qualities because alternatives with high economic quality also have high environmental quality. A value close to minus one shows an extreme need to trade off the qualities: alternatives with high economic quality have low environmental quality and vice versa. The correlation coefficient value of -0.799 in Screen 10.7 indicates a substantial level of trade-off between economic and environmental quality. The ranking of the alternatives will therefore be highly sensitive to changes in the priorities (weights) assigned to economic and environmental quality.

Changing priorities

The diagram can be used to explore the influence of changes in priorities graphically. In the next diagram priority is given to economic quality. This results in a change in slope of the diagonal line.

What to do:

Return to the menu:	**<ESC>**
Increase the **weight** of economic quality to 75:	**<F4> 75 <RETURN>**
Move line:	**<F6>**

This results in the following screen.

Screen 10.8. *Graphical ranking of the alternatives, priority is given to economic quality, year 2030.* Forest and Timber Inquiry Australia.

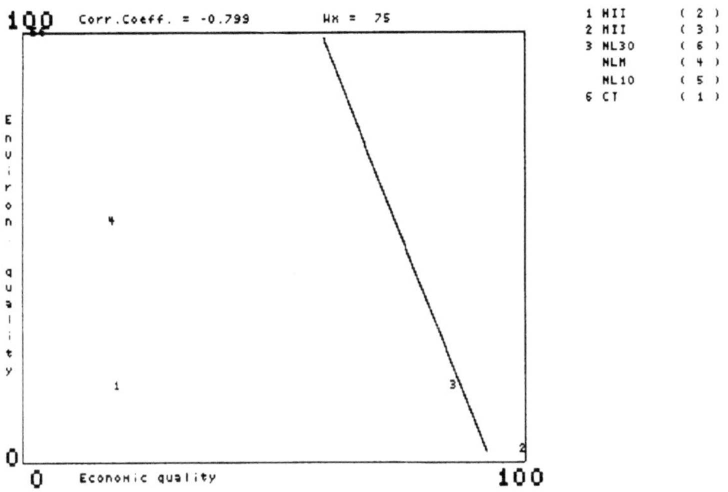

Move line
Move line with cursor keys.
Return to menu with <ESC>

The distance between the intersection and the ideal point is now much smaller for the horizontal axis. This change in slope implies that an option with high economic quality will rank high even if environmental quality is limited. Alternative HII, an alternative with a maximum score on economic quality and a minimum score on environmental quality, is now the first to cross the line; it is followed by alternative MII and, at some distance, NL30 and NLM.

You can now move on to the next module: problem evaluation.

What to do:

Return to the menu:	<ESC>
Continue to the appraisal table:	**<F10>**
Continue:	**<F9>**
Select module C: **problem evaluation**:	**<F2>**

10.4. Problem evaluation

This module contains two main categories of evaluation methods: monetary evaluation methods and multicriteria methods. Since the effects in this problem are measured partly on a ratio scale and partly on a qualitative (---/+++) scale, the Expected Value method has been selected as the most appropriate evaluation method for this problem (see C1 method information for more information on the characteristics of the available methods).
The path this example takes through module C is emphasized in Figure 10.4.

What to do:

Select **Expected Value method**:	**<F5>**
Select **weights**:	**<F2>**
Select **Expected Value method**:	**<F1>**
Select **make**:	**<F1>**
Select the most important effect:	**<F1>**
Select the next important effect:	**<F1>**
Select **as important** and confirm:	**<F2> <ESC>**
Do not add information:	**N**
Execute:	**<F1>**

What to do:

Save the result:	**<F4>**
Select yes and add your own information:	**Y YOURINFO <RETURN>**

The information lines are important to be able to select the required results for sensitivity analysis in the next module. You will now generate one more ranking giving priority to environmental quality and then proceed to sensitivity analysis.

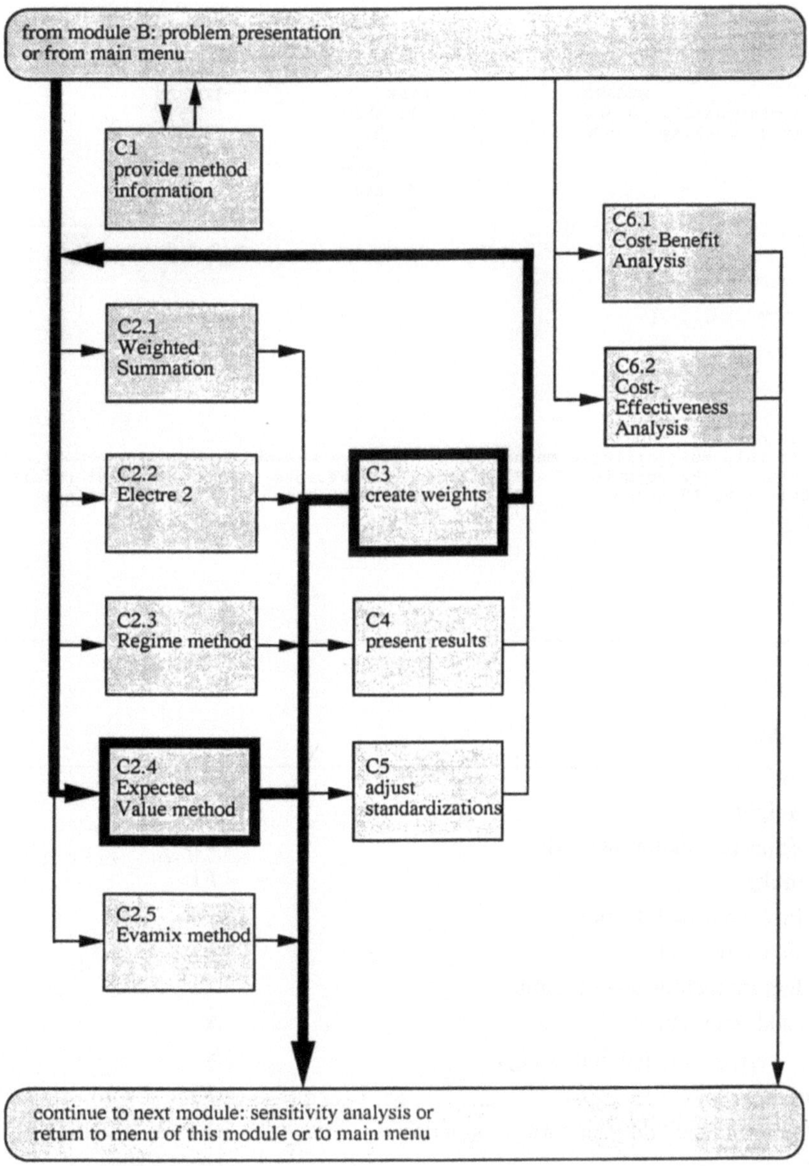

Figure 10.4. Problem evaluation. Forest and Timber Inquiry Australia.

Screen 10.9. Results of the Expected Value method.

Forest and Timber Inquiry Australia.

```
========================Expected Value Meth.==========================

-----------WEIGHTS---------                ----------RANKING--------
RANKING                  SCORES            RANK                 SCORES
 1: Economic quality   0.500                1: NL30              0.51
    Environ. quality   0.500                   MII               0.50
                                               HII               0.50
                                               NL10              0.50
                                            5: NLM               0.35
                                            6: CT                0.16




Select an option:
=Execute this multicriteria method==============================C2.4==
F1.execute      F2.weights      F3.standardize  F4.save results F5.print result
F6.effects tab. F7.graph        F8.other MCA    F9.help         F10.continue
```

What to do:

Select **weights**:	**<F2>**
Select **Expected Value method**:	**<F1>**
Select **make**:	**<F1>**
Select Environmental quality:	**<F2>**
Select Economic quality:	**<F1>**
Select **less important** and confirm:	**<F1><ESC>**
Do not add information:	**N**
Do not overwrite the previous weights:	**N**
Execute:	**<F1>**
Save the result and add your own information:	**<F4> Y YOURINFO <RETURN>**
Do not overwrite the previous results:	**N**
Continue:	**<F10>**
Select **clean-up** to have a look at the results of this module:	**<F4>**
Continue:	**<F3>**
Go to module D: **sensitivity analysis**:	**<F2>**

10.5. Sensitivity analysis

Problem evaluation is used to rank the six alternatives for national forest and timber use. It is shown, for example, that if priority is given to economic quality, alternative HII is the most preferred alternative and alternative CT is the least preferred. To reach this conclusion, criterion scores (included in the effects table), together with criterion weights, were included in the method. If the criterion scores and the criterion weights were completely certain, this conclusion would also be completely certain and sensitivity analysis would not be necessary. Since projections always have a margin of error, and since weights usually represent a range rather than a precise value, the rankings are not completely certain.

In this example we will explore:

- changes in the scores of the economic and environmental quality criteria (effects uncertainty);
- changes in the weights assigned to individual economic and environmental quality criteria (weight intervals);

The path this example takes through module D is emphasized in Figure 10.5.

The sensitivity of alternatives to overall uncertainty in criterion scores can be analyzed using a Monte Carlo approach (Janssen, 1992). An estimate must be provided of the extent (as a percentage) to which actual values may differ from values included in the effects table. A random generator is used to translate this information into a large number of effects tables around the original effects table. Rankings are then determined for all these effects tables. Variations in these rankings are used to determine the sensitivity of the original ranking to uncertainties in effect scores. In this application the scores were varied between plus and minus 25 percent of the scores presented in the effects table.

What to do:

Select **effects uncertainty**:	**<F2>**
Read the information lines:	**<SPACE>**
Select the result with environmental quality = 0.750:	**<SPACE><RETURN>**
Select **calculate**:	**<F1>**
Assign the uncertainty percentage to Value added:	**25 <RETURN>**
Assign the same percentage to the other criteria:	**25 <RETURN>**

This results in the following screen.

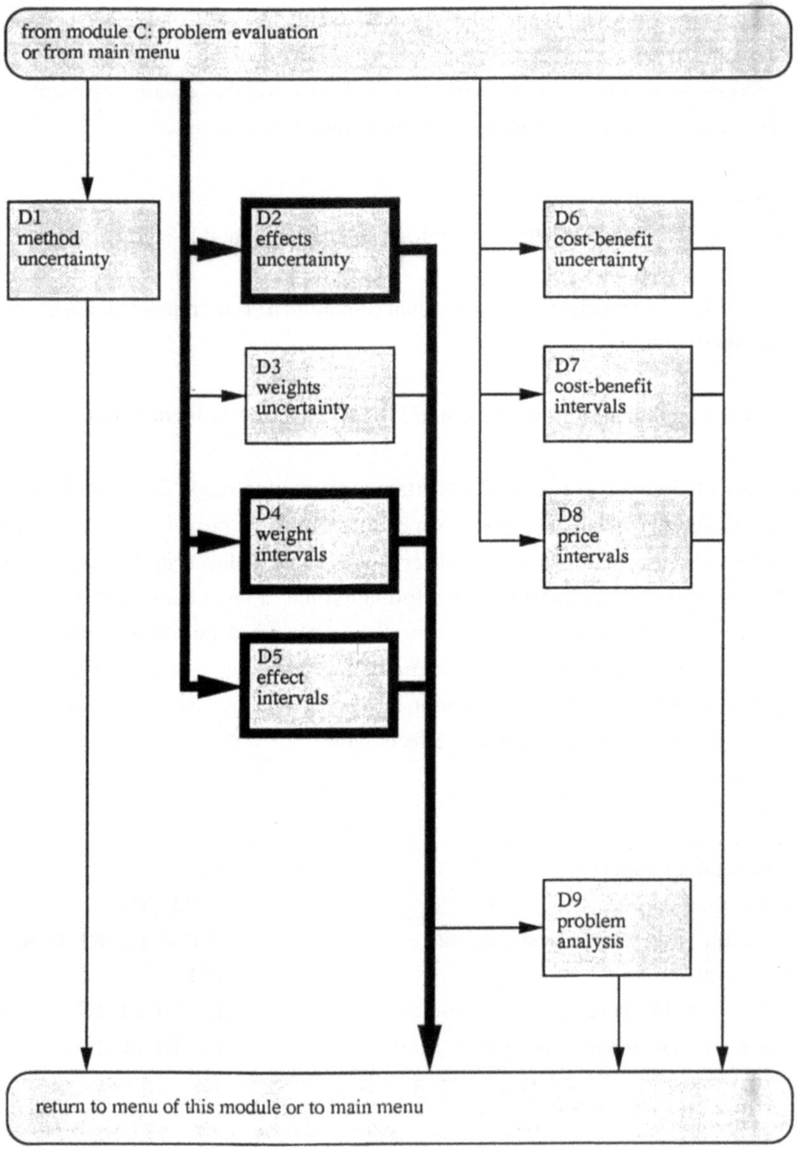

Figure 10.5. Sensitivity analysis. Forest and Timber Inquiry Australia.

Screen 10.10. Results of effects uncertainty. Forest and Timber Inquiry Australia.

probabilities	NL30	NL10	NLM	MII	HII	CT
rank 1	0.43	0.52	0.03	0.01		
rank 2	0.40	0.30	0.11	0.14	0.03	0.02
rank 3	0.13	0.15	0.33	0.23	0.07	0.08
rank 4	0.04	0.01	0.29	0.30	0.20	0.17
rank 5		0.01	0.17	0.17	0.36	0.28
rank 6			0.07	0.16	0.33	0.45
total score	5.22	5.27	3.33	3.08	2.08	1.94

```
Select an option:
==Display conclusion====================================================D2  ==
F1.conclusion   F2.save          F3.print        F4.percentages  F5.information
F6.help         F7.other perc.   F8.other mca    F9.other anal.  F10.continue
```

What to do:

Select **conclusion**:	<F1>

The conclusion shows which part of the rankings is still sufficiently certain given the estimated chance of the criterion scores changing by plus or minus 25 percent. The ? between two alternatives indicates that the rank order of two alternatives is not certain enough. Note that it is not possible to discriminate between NL30 and NL10. It can be concluded, however, that these two alternatives are preferred to all other alternatives. As an example of the sensitivity of the rankings to changes in specific scores, the rankings of alternatives NL30 and NL10 are tested against changes in the Value-added score of alternative NL10. This shows that a small change in the Value added score of NL10 is sufficient to rank alternative NL10 before alternative NL30.

What to do:

Select **other analysis**:	<**F9**>
Select **effect intervals**:	<**F5**>
Read the information lines:	<**SPACE**>
Select the result with economic quality = 0.500:	<**RETURN**>
Select **calculate**:	<**F1**>
Select the alternatives **NL10** and **NL30**:	<**F5**><**F6**>
Select the score of Value added of NL10 (4473.00):	<**CURSOR**><**RETURN**>

This results in the following screen.

Screen 10.11. Results of effect intervals. Forest and Timber Inquiry Australia.

```
=============================== EFFECT INTERVAL =================================

      max ------
          ------                          MCA method      : Expected Value Meth.
          ------                          Weight method   : Expected Value
          ------
          ------
          ------ NL10     > NL30
          ------
          ------                          The score interval is calculated
          ------                          for effect           : Value added
          reverse 4500.96                 of alternative       : NL10
          ++++++                          with starting point :  4473.00
  4473.00 start  NL30     > NL10
          ++++++                          A rank reversal takes place
          ++++++                          at value :   4500.96
          ++++++
          ++++++
          ++++++
        0 ++++++

Select an option:
==Save this combination=========================================================D5  ==
F1.save          F2.print          F3.information  F4.help          F5.other score
F6.other alter.  F7.other MCA      F8.other anal.  F9.continue
```

The correlation coefficient calculated in conjunction with Screen 10.7 already indicates that the rankings will be highly sensitive to changes in the weights assigned to the economic and environmental quality indices. The next screen shows how a change in the relative weights of both indices influences the top two positions in the ranking. Note that, since the weights of both indices add up to one, the weight of environmental quality can be found by subtracting from 1 the weight of economic quality.

What to do:

Select **other analysis**:	**<F8>**
Select **weight intervals**:	**<F4>**
Select the result with economic quality = 0.500:	**<RETURN>**
Select **interval**:	**<F1>**
Select the alternatives **NL30** and **MII**:	**<F6><F3>**
Select the weight of **economic quality** to be analyzed:	**<F1>**

This results in the following screen.

Screen 10.12. Results of weight intervals. Forest and Timber Inquiry Australia.

```
═══════════════════════ WEIGHT INTERVAL ═══════════════════════

      1 ------
        ------                      MCA method      : Expected Value Meth.
        ------                      Weight method   : Expected Value
        ------
        ------ MII      > NL30
        ------
        ------
        reverse   0.503             The weight interval is calculated
  0.500 start  NL30     > MII       for effect        : Economic quality
        ++++++                      with rank number :   1
        ++++++                      and weight        :   0.500
        ++++++
        ++++++                      A rank reversal takes place
        ++++++                      at value :     0.503
        ++++++
        ++++++
        ++++++
      0 ++++++

Select an option:
═Save this interval═══════════════════════════════════════════D4  ══
F1.save          F2.print      F3.information  F4.help        F5.other weight
F6.other alter. F7.other MCA   F8.other anal.  F9.continue
```

Screen 10.12 shows that the position of the two highest ranked alternatives is very sensitive to small changes in the weights around the equal weights point. The changes in weights resulting in rank reversal are so small it is probably better to conclude that, if equal weight is given to both indices, alternatives MII and NL30 rank equal.

Only part of the analysis has been shown in this chapter. The full study included not only the year 2030 but also the year 2000 and a comparison of the relative development of the scenarios between these years.

Multicriteria analysis can be used to select the best option from a set of alternatives, to rank all alternatives from the most preferred to the least preferred, and to improve understanding of the characteristics of an evaluation problem. In the Inquiry's application, the last function proved to be of greater importance than the first two functions. Determining the most preferred alternative, or the ranking between alternatives in this analysis depended critically on how economic or environmental quality was weighted.

The multicriteria analysis did increase understanding of the usefulness of the criteria used to compare the alternatives. But most of all it improved the understanding of the characteristics of the total set of available alternatives in the short and medium term.

You have now completed the first example. You can either continue with one of the other examples or exit the program.

1. What to do to leave the program:

Select **continue**:	**<F9>**
Select **main menu**:	**<F2>**
Select **exit the program**:	**<F7>**

You can now switch off your computer

2. What to do to continue to the next example:

Select **continue**:	**<F9>**
Select **main menu**:	**<F2>**
Select **utilities**:	**<F6>**

This results in the following screen.

Screen 10.13. Utilities. Forest and Timber Inquiry Australia.

```
Current session name  : forest1

Current presentation  : intermediate and final results
Information screens   : on
Appraisal table       : standardized
                        available sessions
     ENERGY          EXAMPLE        FOREST          FOREST1
     NUCLEAR         SOIL

Select an option:
==Change session name==========================================T
F1.change        F2.delete       F3.presentation F4.backup       F5.update
F6.information   F7.appr. table  F8.continue
```

What to do:

Change to a new session:	**<F1>**
Enter a new session name:	**NEW_NAME <RETURN>**
Answer yes to use the results of a previous session:	**Y**
Select the example you want to work with:	**<CURSOR><RETURN>**
Continue to the main menu:	**<F8>**

Continue to the chapter that describes the example you have selected.

11. SOIL CONSERVATION IN MAPHUTSENG, LESOTHO

11.1. Introduction

The description of the Farm Improvement with Soil Conservation (FISC) project in Lesotho included in this chapter is a simplified version of the study published in Bojö (1989).[1] Further work has since been completed on this subject and is published in Bojö (1991).[2]

The kingdom of Lesotho is completely surrounded by the Republic of South Africa. Lesotho is a highland country. Its lowest point of elevation is 1300 m, which is the highest of any nation.

The Farm Improvement with Soil Conservation (FISC) project is located in Mohale's Hoek District in the south of Lesotho (see Figure 11.1). The land use in the area is primarily cultivation and grazing. The pressure on the land is very high and even clearly unsuitable land is cultivated. The overriding aim of the FISC project is to raise agricultural production.

The project has rehabilitated old terrace structures and constructed new ones and added stone-paved waterways and cut-off drains. It has promoted hybrid maize and sorghum, fodder grasses, tree planting for fruit and fuel wood and rotational grazing on communal land.

The aim of this economic analysis is to determine what surplus society could be expected to gain over the life-time of the project's impact. This is captured by the Net present value. The yield of the resources invested in the FISC project could also be compared with alternative investment opportunities. This can be done explicitly by calculating the Internal rate of return (IRR) of the project.

This application includes three basic steps.

---

1 Bojö, J. (1989). Case study: Benefit-Cost analysis of soil conservation in Maphutseng, Lesotho. In: Dixon, J.A., D.E. James, and P. Sherman, The economics of dryland management, Earthscan, London.

2 Bojö, J. (1991). *The economics of land degradation: theory and applications to Lesotho.* Stockholm School of Economics, Stockholm. This publication can be ordered from the Stockholm School of Economics, Box 6501, S-11383, Stockholm, Sweden.

The authors would like to thank Dr. Bojö for the opportunity to use his study, even though this limited example does not do full credit to his work.

Figure 11.1. Regions of Lesotho (Source: Bojö, 1991).

Problem definition
In this step the discounted effects table is created. The various costs and benefits are identified and estimated for the projected life time of the project. Costs and benefits are discounted to a base year.

Problem presentation
In the second step a price is attached to non-monetary benefits and costs. The effects table is transformed into a cost-benefit sheet.

Problem evaluation
The net present value, the benefit-cost ratio and the internal rate of return are determined in the third and last step.

What to do:

Go to the DEFINITE directory:	**CD DEFINITE**	**<RETURN>**
Start the program:	**DEFINITE**	**<RETURN>**
Enter your own session name:	**YOURNAME**	**<RETURN>**
Answer yes to the question:	**Y**	
Select the example **Soil**:	**<CURSOR><CURSOR><CURSOR><CURSOR>**	**<RETURN>**

It may take some time to go through the whole example. Return to the main menu and exit the program if you want to stop before reaching the end of the example. When you start the example again you can continue where you stopped.

11.2. Problem definition

As stated above, the aim of this economic analysis is to determine what surplus society could be expected to gain over the life-time of the project's impact. The effects table therefore only includes one alternative to be evaluated. Implicitly this alternative is evaluated against the no-project alternative. In this module the costs and benefits are first identified and estimated for the projected life-time of the project. Costs and benefits are discounted to a base year. The path this example takes through module A is emphasized in Figure 11.2.

What to do:

Enter module A: **problem definition:**	**<F1>**
Enter procedure A1: **create an effects table**	**<F1>**

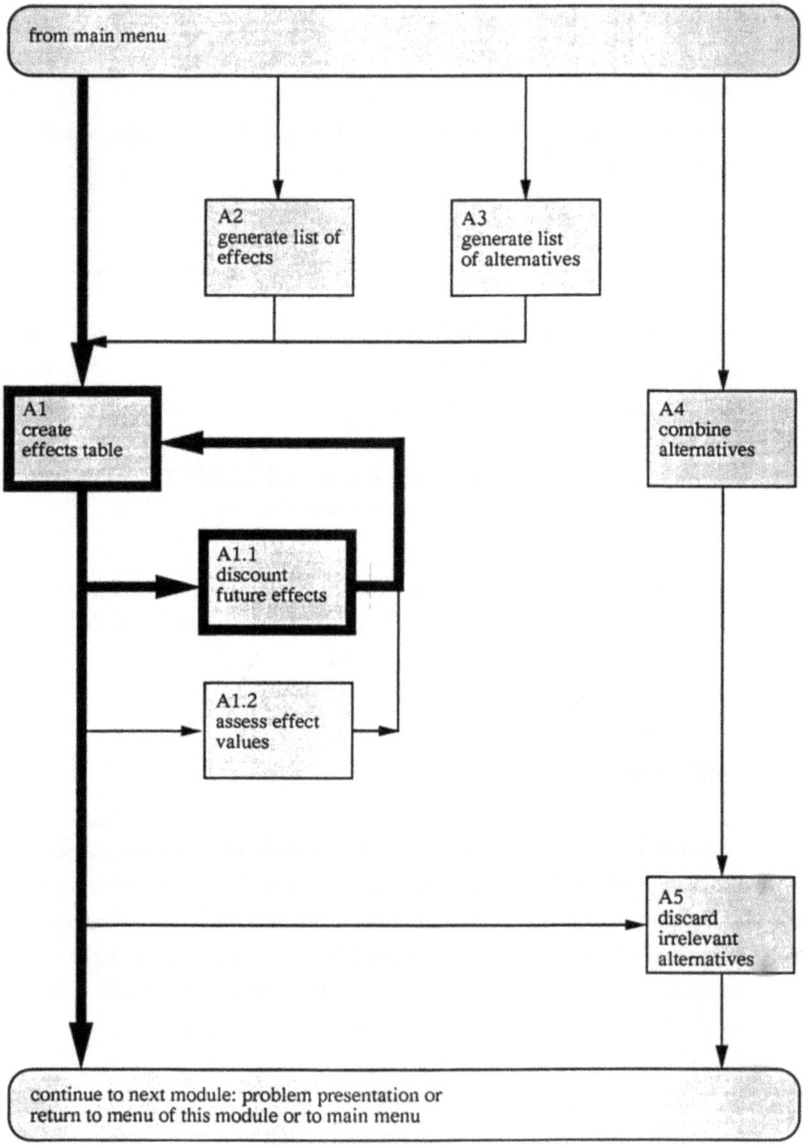

Figure 11.2. Problem definition. Soil conservation in Maphutseng.

Information screens are presented upon entering each procedure. Hit **<ANY KEY>** to continue. The instruction to hit any key is not included in the "what to do: boxes".

Screen 11.1. Effects table. Soil conservation in Maphutseng.

```
gedisconteerd        Project
=======================================================

=======================================================
Mais                    515.61
Sorghum                 329.65
Fodder                  446.84
Cons.Mais               143.35
Cons.Sorghum             84.09
Fuel                    303.04
Residual                1.6E+04
Project costs          -4.7E+05
Input costs            -1.7E+05
Crop loss              -3.1E+04



Select an option:
==Add an alternative===================================A1  ==
F1.alternative  F2.effect       F3.descriptor  F4.edit      F5.units
F6.time         F7.information  F8.help        F9.continue  F10.more
```

The effects table shows the benefits in terms of the increased production of maize, sorghum, fodder and fuel wood. Conservation maize and conservation sorghum refer to benefits that result for those part of the project where the project only included soil conservation. Increased production is measured in tonnes per year, fuel wood in m^3 per year. Residual benefits, Project costs and Input costs and Crop loss are measured in Maloti.

The values shown in the effects table are discounted values. The project was started in 1985 and costs and benefits are estimated for the following 20 years. To be able to determine the discounted effects table costs and benefits need to be specified for each of these years.

What to do:

Select **time**:	**<F6>**
Select **next year**:	**<F2>**
Select **time.003** (information = year 1986):	**<F3>**

This results in the following screen.

Screen 11.2. Effects table 1986. Soil conservation in Maphutseng.

```
003: 1986          Project
=====================================================================

=====================================================================
Mais                 48.00
Sorghum              32.00
Fodder               12.00
Cons.Mais             2.00
Cons.Sorghum          1.00
Fuel
Residual
Project costs     -1.2E+05
Input costs       -1920.00
Crop loss         -1164.00




==Edit next year===============================================A1.1==
F1.edit          F2.next year    F3.previous    F4.information  F5.help
F6.change year   F7.units        F8.continue
```

This table shows the costs and benefits that occur in the first year of the project 1986. Note that benefits are still relatively low and costs are high. The trees planted for fuel wood do not yet provide any benefit.

Costs and benefits that occur during a certain year are attributed to the end of that year. Costs and benefits occurring in the base year are not discounted. This implies that the investment decision is assumed to take place at the first of January of year 1. It is assumed implicitly that costs and benefits in year 0 occur at the end of this year. Costs in year 0 are usually investment costs and there are usually no benefits. If the investment costs are spread over a longer period it can be preferable to also discount the investment costs. The year of investment should then be entered as year 1 and the base year as the year before. In this case the effects table of the base year should be empty. In this application both costs and benefits of the first year are discounted. Therefore the base year is 1984 and the first project year is 1985 (see Figure 11.3).

Costs and benefits have been determined for all 20 years and can be viewed by selecting next year. Next, all 20 effects tables are transformed into a single discounted effects table that includes discounted values for the common base year. In this example both discount rates and price indices will be specified to accommodate changes in prices over time.

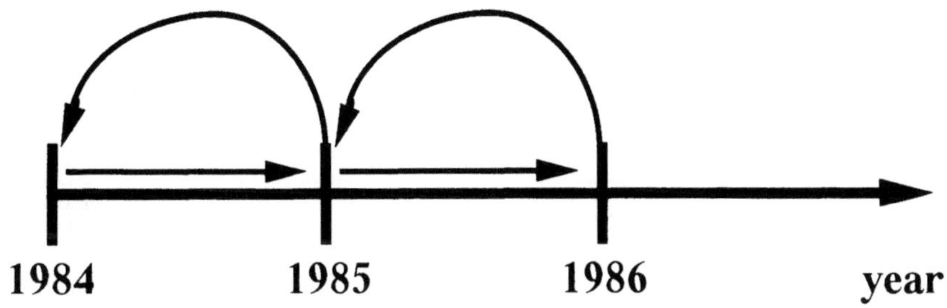

Figure 11.3. Discounting to the base year.

What to do:

Continue: **<F8>**
Select option 3 discount rate and price indices: **3 <RETURN>**
Specificy discount rates and price indices according to the screen below

---

A discount rate of 5 percent is specified for all costs and benefits. This discount rate is an estimate of the opportunity cost of capital in Lesotho. All calculations are made in real, not nominal prices. However, it cannot be expected that all relative real prices will remain the same for the duration of the project. The price indices can be used to correct for changes in real prices. The price index reflects the changes in real prices over time. In this study the real prices of maize and sorghum were assumed to decrease at a rate of 2 percent per year. Constant real prices were assumed for Fodder and Fuel. The monetary effects do not require a price or price index. The cost-benefit sheet is specified in the next module. Note that the price index needs to be specified here but the price itself is not specified until the cost-benefit sheet is specified.

Screen 11.3. Discount rates and price indices. Soil conservation in Maphutseng.

discounted	DiscRate	pr index	UNITS
Mais	5.00	-2.00	ton/year
Sorghum	5.00	-2.00	ton/year
Fodder	5.00		ton/year
Cons.Mais	5.00	-2.00	ton/year
Cons.Sorghum	5.00	-2.00	ton/year
Fuel	5.00		M3/year
Residual	5.00		x 1
Project costs	5.00		x 1
Input costs	5.00		x 1
Crop loss	5.00		x 1

A1.1

F1.edit F2.calculate F3.information F4.help

What to do:

Select **calculate**: **<F2>**

The effects table is the final result of problem definition. You can now continue to the next module: problem presentation. This results in the discounted effects table as was shown in Screen 11.1.

11.3. Problem presentation

In this module the discounted effects table is transformed into a cost-benefit sheet. The path this example takes through module B is emphasized in Figure 11.4.

What to do:

Continue:	**<F5>**
Go to **module B**:	**<F2>**
Go to **cost-benefit sheet**:	**<F4>**

This results in the following screen.

Screen 11.4. Cost-benefit table. Soil conservation in Maphutseng.

cost-benefit		Units	Prices	Project
Effects	Cost Benefit		x 1	
Project costs	C	x 1	1	4.7E+05
Input costs	C	x 1	1	1.7E+05
Crop loss	C	x 1	1	3.1E+04
Residual	B	x 1	1	1.6E+04
Mais	B	ton/year	?	515.61
Sorghum	B	ton/year	?	329.65
Fodder	B	ton/year	?	446.84
Cons.Mais	B	ton/year	?	143.35
Cons.Sorghum	B	ton/year	?	84.09
Fuel	B	M3/year	?	303.04

```
Select an option:
=Assign a price to an effect==================================B5  ==
F1.prices       F2.time        F3.neg.benefits F4.move       F5.C/B sheet
F6.graph        F7.print       F8.information  F9.help       F10.continue
```

Screen 11.4 shows three monetary costs and one monetary benefit. These four effects do not need to be priced since they are already measured in monetary units, in this case in Maloti. The six remaining benefits are measured in physical units and no price is yet available to value these benefits.

In this economic analysis, the relevant prices for increased output of crops are the economic value (to the economy) rather than the financial value (to the farmer). The economic value to Lesotho of one extra tonne of maize equals the import cost saved. This has been calculated as 380 Maloti per tonne of maize in 1986 prices. For Sorghum, the corresponding figure is 370 Maloti. Since 1984 has been selected in this example as the base year and a downward trend in the real prices of maize and sorghum has been assumed these prices are adjusted to 395.92 (maize) and 384.7 (sorghum) before they are entered in the cost-benefit table. A constant real price of 4.50 Maloti per bag of 20 kg (225 Maloti/ton) was assumed for fodder. Since additional fuel wood will to some extent replace alternative sources of energy a substitution value is estimated. The total substitution value equals 28.71 Maloti/m^3.

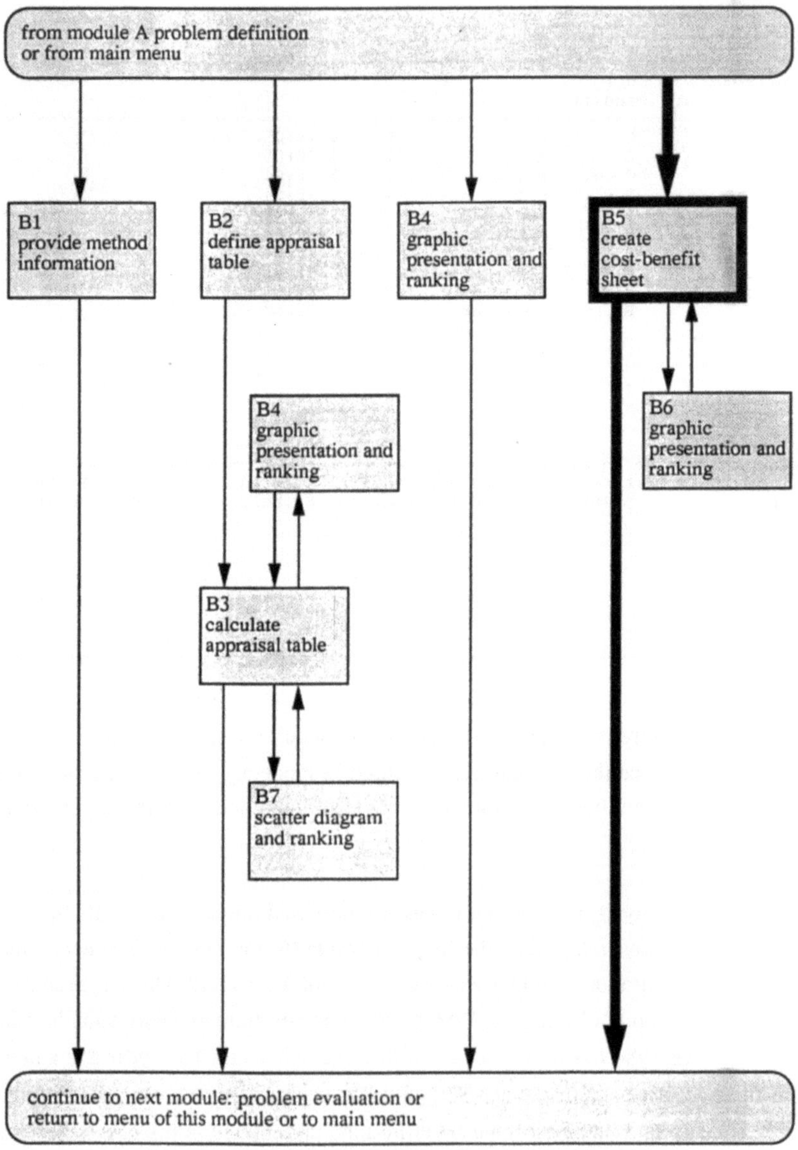

Figure 11.4. Problem presentation. Soil conservation in Maphutseng.

You can now enter these prices in the cost-benefit table. After entering the prices this table is transformed into a monetary cost-benefit table (Screen 11.5) and a cost-benefit sheet (Screen 11.6).

What to do:

Enter a price for Maize:	**<F1><CURSOR>**	**395.92**	**<RETURN>**
Enter a price for Sorghum:	**<CURSOR>**	**384.70**	**<RETURN>**
Enter a price for Fodder:	**<CURSOR>**	**225.00**	**<RETURN>**
Enter a price for Maize:	**<CURSOR>**	**395.92**	**<RETURN>**
Enter a price for Sorghum:	**<CURSOR>**	**384.70**	**<RETURN>**
Enter a price for Fuel wood:	**<CURSOR>**	**28.71**	**<RETURN>**
Go to **cost-benefit sheet**:	**<F5>**		

This results in a monetary cost-benefit table (all scores in Maloti).

Screen 11.5. Monetary cost-benefit table. Soil Conservation in Maphutseng.

Mon. C/B table		**Units**	**Prices**	**Project**
Effects	**Cost Benefit**		**x 1**	
Project costs	C	x 1	1	4.7E+05
Input costs	C	x 1	1	1.7E+05
Crop loss	C	x 1	1	3.1E+04
Residual	B	x 1	1	1.6E+04
Mais	B	ton/year	395.92	2.0E+05
Sorghum	B	ton/year	384.70	1.3E+05
Fodder	B	ton/year	225.00	1.0E+05
Cons.Mais	B	ton/year	395.92	5.7E+04
Cons.Sorghum	B	ton/year	384.70	3.2E+04
Fuel	B	M3/year	28.71	8700.28

This is the cost-benefit table with monetary values.
═Display as a cost-benefit sheet═══════════════════B5 ═
F1.continue F2.print

What to do:

Continue:	**<F1>**

This results in the following screen.

Screen 11.6. Cost-benefit sheet. Soil conservation in Maphutseng.

```
              Project
               x 1
        Costs                  Benefits

Projectc   4.7E+05     Residual    1.6E+04
Inputcos   1.7E+05     Mais        2.0E+05
Croploss   3.1E+04     Sorghum     1.3E+05
                       Fodder      1.0E+05
                       Cons.Mai    5.7E+04
                       Cons.Sor    3.2E+04
                       Fuel        8700.28

Select an option:
=Select a different alternative=========================B5 =
F1.alternative  F2.right/left   F3.print       F4.continue
```

11.4. Problem evaluation

The aim of this economic analysis is to determine what surplus society could be expected to gain over the life-time of the projects impact. This is captured by the net present value. The yield of the resources invested in the FISC project could also be compared with alternative investment opportunities. This is done implicitly because a real social opportunity of capital discount rate was used. This can also be done explicitly by calculating the internal rate of return (IRR) of the project.

The net present value, the benefit-cost ratio and the internal rate of return are determined in this third and last step. The path this example takes through module C is emphasized in Figure 11.5.

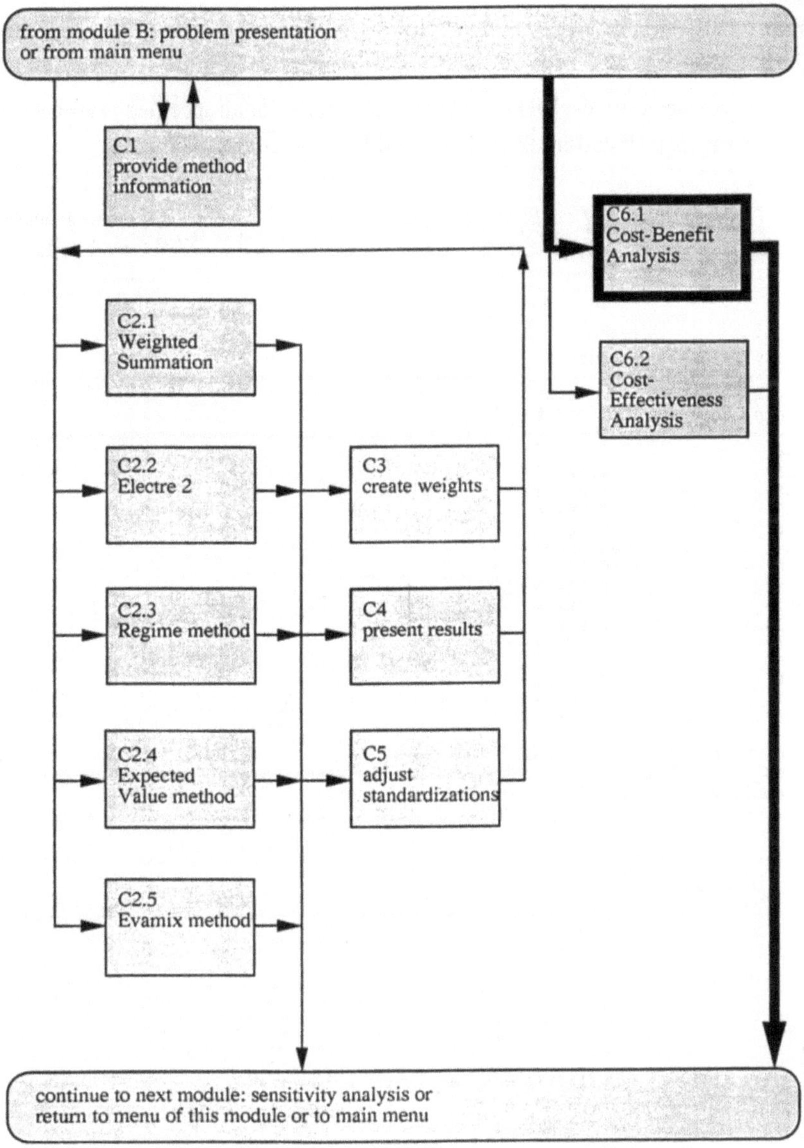

Figure 11.5. Problem evaluation. Soil conservation in Maphutseng.

What to do:

Return to the previous screen:	<F4>
Continue:	<F10>
Go to **module** C:	<F2>
Select **Cost-Benefit analysis**	<F1>
Continue:	<F1>

As an intermediate result the balances of cost and benefits for all specified years are presented. This table shows that starting from 1991 total benefits exceed total costs.

Screen 11.7. Balances of costs and benefits per year. Soil conservation in Maphutseng.

```
Yearly balances   Project
================================================================
Benefit/Cost         0.81
Intern. Rate         1.2 %
================================================================
1984
1985              -2.0E+05
1986              -9.2E+04
1987              -4.7E+04
1988              -3.7E+04
1989              -2.0E+04
1990              -9528.36
1991               4.8E+04
1992               2.7E+04
1993               2.8E+04
1994               2.9E+04
1995               3.0E+04
1996               3.0E+04
1997               3.1E+04
1998               4.9E+04

Examine the table using the cursor keys
==End this procedure and continue==============================C6.1==
F1.continue     F2.print
```

As the final results[1] of Cost-Benefit analysis, the net present value, the benefit-cost ratio and the internal rate of return are calculated.

What to do:

Continue:	<F1><F1>

Screen 11.8. Results Cost-Benefit analysis. Soil conservation in Maphutseng.

Cost-Benefit Analysis

Alternatives Disc.rate= 5.00 %	Net Present Value	Benefits/Costs	Internal Rate of Return (%)
Project	-1.3E+05	0.81	1.16

```
Select an option:
==Sort according to Net Present Value==================================C6.1==
F1.sort NPV     F2.sort B/C     F3.sort IRR     F4.print     F5.information
F6.graph NPV    F7.graph B/C    F8.graph IRR    F9.help      F10.continue
```

1 The equation used to discount future costs and benefits is shown below.

$$x_{ji}^{d} = \sum_{t^0}^{t^n} x_{ji}^{t} * \frac{(1+p_i/100)^{t-t^0}}{(1+d_i/100)^{t-t^0}}$$

In this equation x_{ji}^d represents the discounted effect i of alternative j and x_{ji}^t represents this effect at year t. The base year is shown as t^0 and the last year to be included (the planning horizon) as t^n. In this equation r_i represents the discount rate and as p_i the price index, both entered as percentages. The price index can be used to correct for changes in prices over time. The discount rate can be specified as a uniform discount rate for all effects or can be specified for each effect separately.

 180

These results show that, given all assumptions made, the project makes a net present loss that is significant in comparison to the resources invested. This is also reflected in a benefit-cost ratio that is below 1 and an Internal rate of return that is below the estimated opportunity cost of capital of 5 percent.

The project has therefore not demonstrated success in relation to its goal of increasing the production on small scale farming units when this increase in production is put in relation to the cost of the project. The project uses simple, labour-intensive techniques. It is based on consultation and people's participation. It therefore has several positive features associated with a successful conservation project. Not all of these positive features are included as monetary benefits. Explanations are also to be found in the very difficult environment of low expected returns to agricultural investments, coupled with high risk exposures and low incomes among farmers.

You have now completed the second example. You can either continue with one of the other examples or exit the program.

1. What to do to leave the program:

Select **continue**:	**<F10>**
Go to the **main menu**:	**<F3>**
Exit the program:	**<F7>**

You can now switch off your computer

2. What to do to continue to the next example:

Select **continue**:	**<F10>**
Go to the **main menu**:	**<F3>**
Select **utilities**:	**<F6>**
Change to a new session:	**<F1>**
Enter a new session name:	**NEW_NAME <RETURN>**
Answer yes to use the results of a previous session:	**Y**
Select the example you want to work with:	**<CURSOR><RETURN>**
Continue to the main menu:	**<F8>**

Continue to the chapter that describes the example you have selected.

12. ENVIRONMENTAL EFFECTS OF ELECTRICITY PRODUCTION IN THE NETHERLANDS

12.1. Introduction

Since the mid-seventies debates about energy policy have been intense, particularly on the choice of energy conversion systems for power generation. The focus of these debates shifted from dependence on oil to calamities associated with nuclear power, and from acidifying emissions from fossil fuel combustion to disposal of nuclear waste.

The problem of how best to produce electricity goes through cycles of low and high political importance. At present political importance is high as a result of a renewed discussion on nuclear energy and the emphasis on CO_2 emissions in relation to the greenhouse effect. Allocation of production capacity in the Netherlands is the subject of the Integral Plan on Electricity Production. In this example alternatives for electricity production in the Netherlands are evaluated with respect to the environmental effects of these alternatives.

As an example Figure 12.1 shows the expected acid deposition that results from the maximum nuclear alternative. This figure shows that even with maximum nuclear and extra pollution control the relative modest policy objective of 3000 acid equivalents per hectare per year is exceeded in most regions of the Netherlands.

This example is based on an evaluation study commissioned by the Ministry of the Environment of the Netherlands. The example presented here is a highly simplified representation of this study.[1]

This application includes four steps:

1. **Problem definition**
   In this step the effects table is created. The effects table is made up of the alternatives as columns, and a number of criteria that reflect the predicted effects of the three alternatives as rows.
2. **Problem presentation**
   This step involves aggregating the effects measured in physical scores such as emissions and land use into ten policy categories such as acidification and greenhouse effect.
3. **Problem evaluation**
   The third step involves the use of Weighted Summation to rank the alternatives.

---

[1] Janssen R. (1992). *Multiobjective decision support for environmental management.* Kluwer Academic Publishers, Dordrecht. Chapter 8.

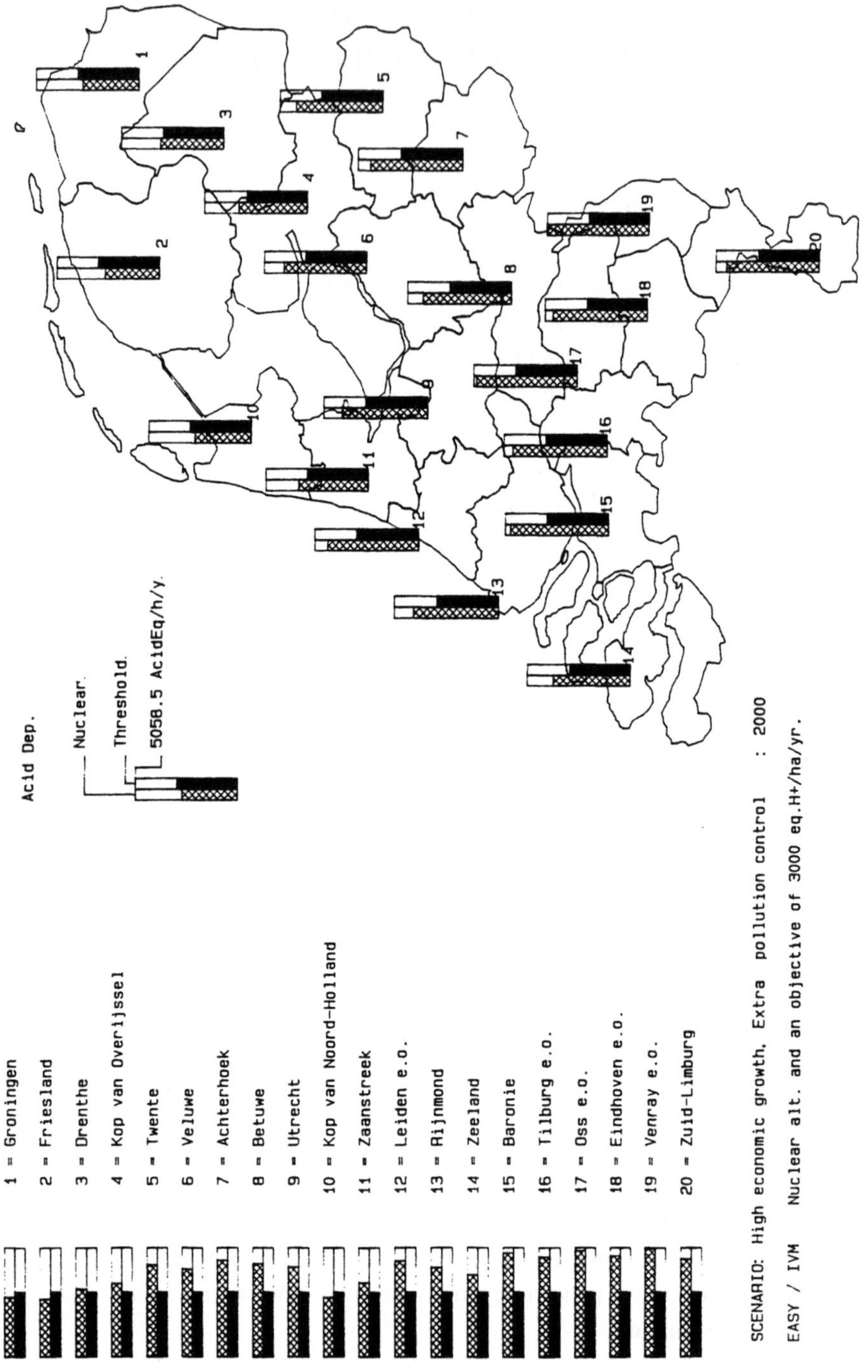

Figure 12.1. *Acid deposition: Nuclear alternative compared with a policy objective of 3000 acid equivalents/hectare/year (scenario: High economic growth, Extra pollution control).*

4. **Sensitivity analysis**
   Weight intervals are determined to analyze the effect of changing weights on the evaluation results.

What to do:

Go to the DEFINITE directory:	**CD DEFINITE**	**<RETURN>**
Start the program:	**DEFINITE**	**<RETURN>**
Enter your own session name:	**YOURNAME**	**<RETURN>**
Answer yes to the question:	**Y**	
Select the example **Energy**:		**<RETURN>**

It may take some time to go through the whole example. Return to the main menu and exit the program if you want to stop before reaching the end of the example. When you start the example again you can continue where you stopped.

12.2. Problem definition

In this example procedure A1 is used to create the effects table. The effects table includes three alternatives: the Nuclear, Coal and Gas alternative, and the direct effects on the environment caused by these alternatives. The path this example takes through module A is emphasized in Figure 12.2.

What to do:

Enter module A: **problem definition:**	**<F1>**
Enter procedure A1: **create an effects table:**	**<F1>**

Information screens are presented upon entering each procedure. Hit **<ANY KEY>** to continue. The instruction to hit any key is not included in the "what to do: boxes".

This results in the following screen.

The effects table includes emissions such as SO_2 and Emissions NO_x and other direct effects such as production of solid waste and land use. Hit **<F5>** to see the measurement units of the effects. The measurement units range from PetaJoule per year (PJ/yr) and KiloTon per year (Kton/yr to decibel (dBa) and hectares (ha). Note that all effects are measured on a ratio scale.

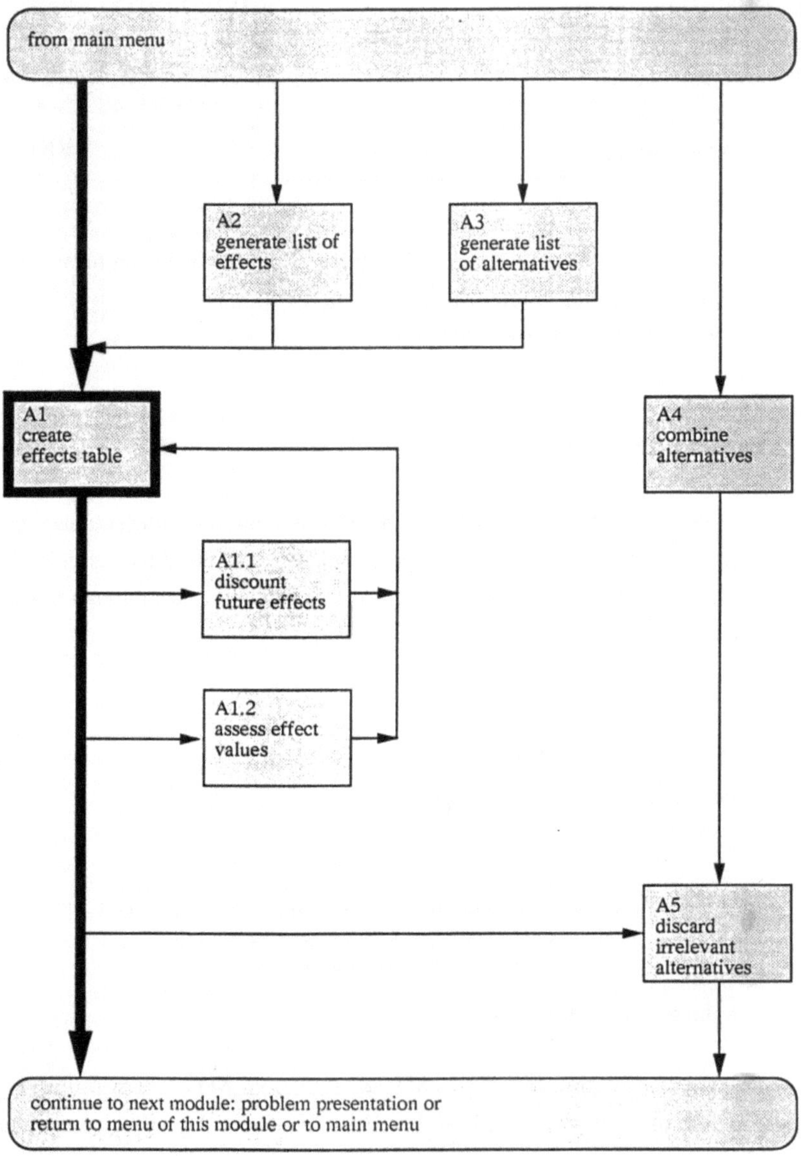

Figure 12.2. Problem definition. Electricity production in the Netherlands.

```
 effects table     Nuclear  Coal      Gas
===========================================================================
Capacity             1.8E+04  1.8E+04  1.6E+04
Production            259.00   271.00   252.00
===========================================================================
SO2                 -3.1E+04 -4.6E+04 -2.4E+04
NOx                 -5.1E+04 -6.9E+04 -5.2E+04
CO2                 -3.1E+07 -4.3E+07 -3.3E+07
CO                  -8342.00 -1.1E+04 -8453.00
CxHy                -3341.00 -3727.00 -4049.00
Particulates        -2862.00 -4209.00 -2164.00
Noise                -417.00  -419.00  -438.00
Radiation           -1.9E+06 -3.7E+05 -3.7E+05
Heat                 -331.00  -313.00  -287.00
Solid Waste         -3.4E+05 -7.7E+06 -1.9E+06
Nuclear Waste       -1400.00  -250.00  -250.00
Landuse             -7139.00 -7148.00 -7055.00
Indirect landuse    -1.8E+05 -6.6E+04 -6.0E+04
Visibility          -1.1E+04 -1.2E+04 -1.0E+04
Area at risk        -4000.00  -800.00  -800.00

Select an option:
=Add an alternative===============================================A1   =
F1.alternative  F2.effect       F3.descriptor   F4.edit       F5.units
F6.time         F7.information  F8.help         F9.continue   F10.more
```

The effects table is the final result of problem definition. You can now continue to the next module: problem presentation.

What to do:

Select **continue** to leave this procedure:	<F9>
Continue to module B: **problem presentation**:	<F2>

12.3. Problem presentation

In this module the effects table is transformed into an appraisal table. An appraisal includes a maximum of 10 appraisal categories. The alternatives are presented and ranked graphically. The path this example takes through module B is emphasized in Figure 12.3.

What to do:

Select **appraisal table**:	<F2>

This results in the following screen.

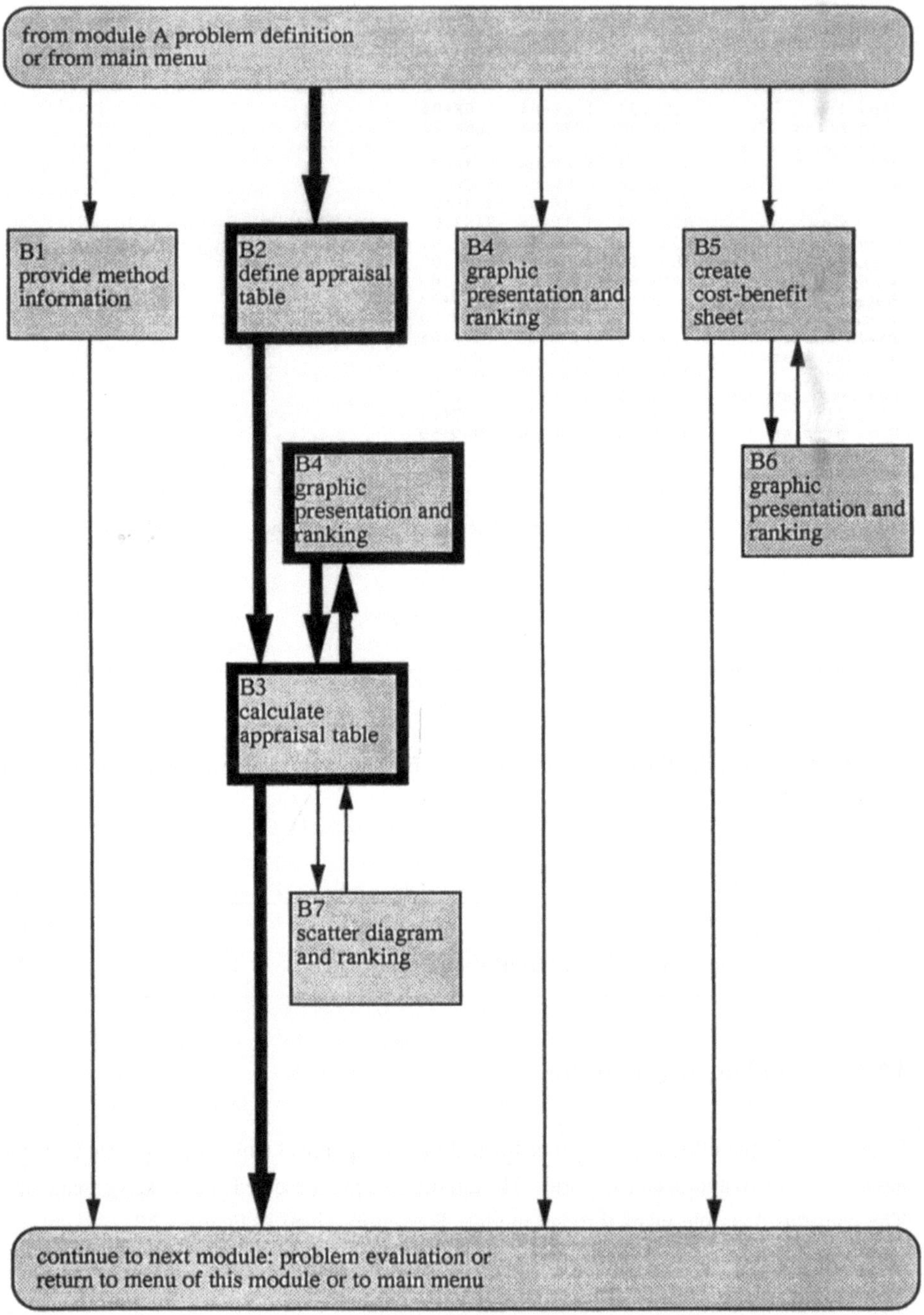

Figure 12.3. Problem presentation. Electricity production in the Netherlands.

Screen 12.2. Definition of the appraisal table. Electricity production in the Netherlands.

definitions	SO2	NOx	CO2	CO	CxHy	Particul	Noise
units	ton/y	ton/y	ton/y	ton/y	ton/y	ton/y	dB(a)
Acidification	31.25	21.74					
Photochemical	0.50				1.00		
Greenhouse			1.00	1.00			
Eutrophication		1.00					
Export Acid	25.94	17.61					
Solid waste							
Landscape							
Land use							
Risk							
Nuclear waste							

```
Select an option:
=Calculate the appraisal table=====================B3 ==
F1.calculate    F2.edit
```

The definitions table includes the policy categories, such as acidification, and the coefficients to calculate the appraisal scores from the effects. In defining the values of the coefficients assumptions are made about the relative contribution of various effects to a policy category: acidification, for example, is defined as the sum of the acidifying capacities of SO_2 and Emissions NO_x.

There are two possibilities to calculate the appraisal scores from the effects:

1. The appraisal scores are calculated as the weighted sums of the effects. This is the default setting;
2. The appraisal scores are calculated as the weighted sums of the standardized effects.

To change the setting of the appraisal table you need to go to utilities from the main menu. In this example the appraisal scores are calculated as the weighted sums of the effects. This is the default setting. Unlike the Forest example the effect scores are not standardized before summation.

What to do:

Calculate the appraisal table:	**<F1>**

This results in the following screen.

Screen 12.3. Appraisal table. Electricity production in the Netherlands.

```
appraisal          Nuclear   Coal      Gas

Acidification      -2.1E+06 -2.9E+06 -1.9E+06
Photochemical      -1.9E+04 -2.7E+04 -1.6E+04
Greenhouse         -3.1E+07 -4.3E+07 -3.3E+07
Eutrophication     -5.1E+04 -6.9E+04 -5.2E+04
Export Acid        -1.7E+06 -2.4E+06 -1.5E+06
Solid waste        -3.4E+05 -7.7E+06 -1.9E+06
Landscape          -1.1E+04 -1.2E+04 -1.0E+04
Land use           -2.0E+05 -8.8E+04 -8.1E+04
Risk               -4.5E-05 -1.5E-05 -1.5E-05
Nuclear waste      -1400.00  -250.00  -250.00


Select an option:
==Edit the definitions table================================================B3  ==
F1.definitions  F2.move           F3.graph         F4.diagram       F5.units
F6.print        F7.information    F8.help          F9.continue
```

This table presents an overview of the relative qualities of the alternatives with respect to the policy categories. Similar to the effects table, this table can be presented as a graph.

What to do:

Select **graph**:	**<F3>**
Rank the alternatives:	**<F4>**

This results in the following screen.

Screen 12.4. Graphic presentation of the appraisal table.

Electricity production in the Netherlands.

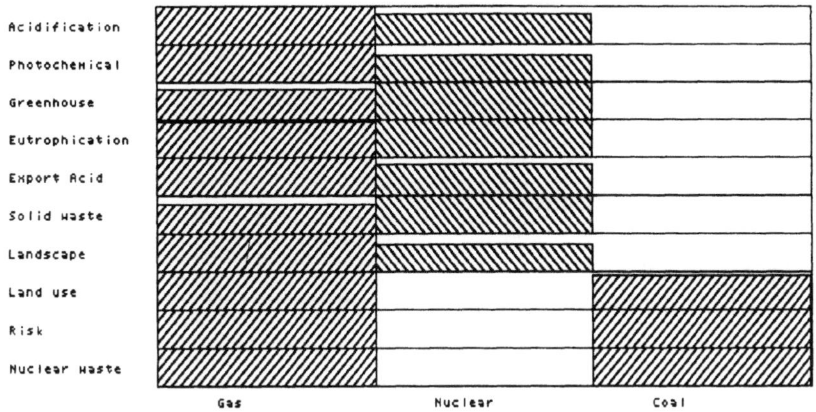

Select an option:
=Rank the alternatives =
F1.exchange row F2.exchange clm F3.print F4.rank F5.information
F6.Help F7.continue

To produce Screen 12.4 the scores are standardized between zero (the worst alternative), and one (the best alternative), for each category. The highest bar represents the best score.

The alternatives are ranked graphically by exchanging rows and columns. In Screen 12.4 the rows are ordered according to the priority assigned to each category, and the columns are ordered in such a way that the blackest ones end up in the upper left corner and the lower right corner of the diagram. DEFINITE performs the sorting of the columns automatically using a combination of the Expected Value method and Weighted Summation.

Screen 12.4 shows that, in ordering the alternatives, a trade-off had to be made between the categories related to nuclear production and the categories related to fossil fuel production. It can also be seen that no trade-off is required between the categories acidification, photochemical pollution and eutrophication; the ranking of the alternatives is not influenced by the relative priorities assigned to these categories.

12.4. Problem evaluation

The appraisal table calculated in module B is the input of the multicriteria methods in problem evaluation. The path this example takes through module C is emphasized in Figure 12.4.

What to do:

Return to the appraisal table:	**<F7>**
Continue:	**<F9>**
Go to module C: **problem evaluation**:	**<F2>**
Select **multicriteria analysis**:	**<F1>**
Select **Weighted Summation**:	**<F2>**

You have now entered Weighted Summation. However, before you can start calculation you need to assign weights to the effects.

What to do:

Select **weights**:	**<F2>**
Select **Expected Value method**:	**<F1>**
Select **make**:	**<F1>**
Select **Acidification**:	**<F1>**
Select **Photochemical**:	**<F1>**
Select **less important**:	**<F1>**
Continue with the other categories	...
Confirm the weights:	**<ESC>**
Add your own information line:	**Y YOURINFO <RETURN>**
Select **execute**:	**<F1>**

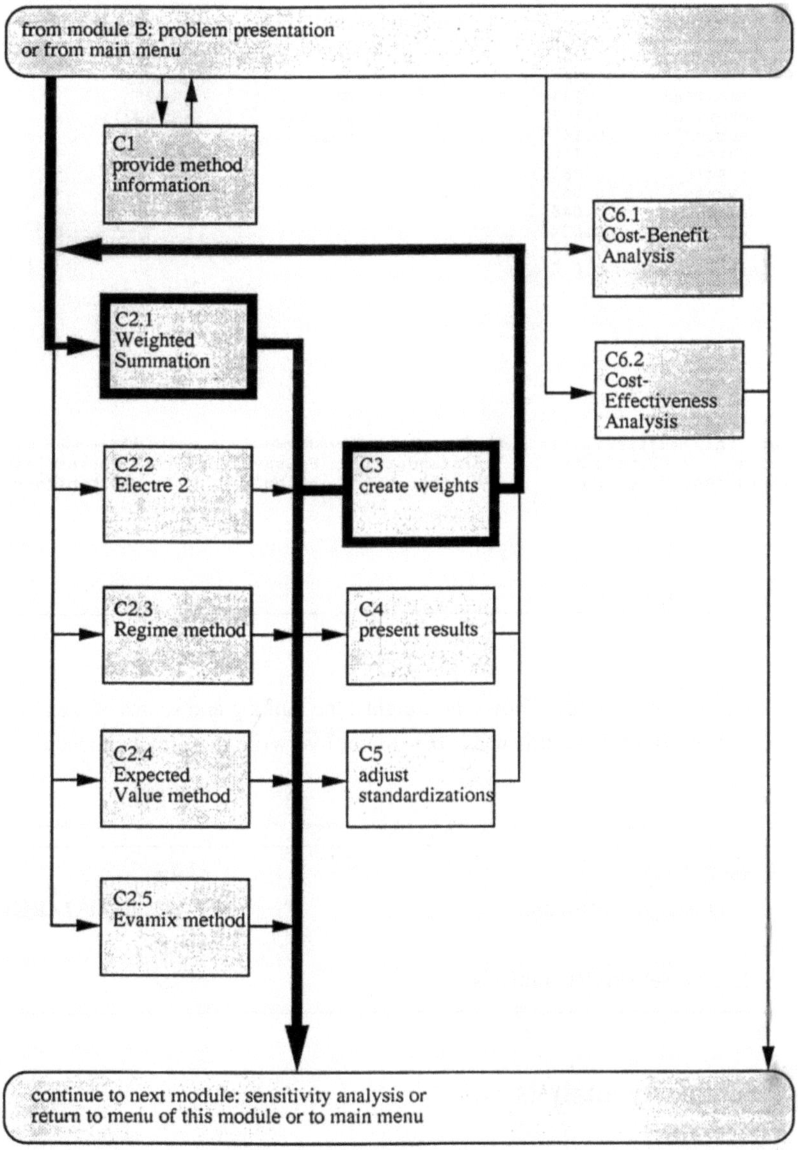

Figure 12.4. Problem evaluation. Electricity production in the Netherlands.

This results in the following screen.

Screen 12.5. Weighted summation. Electricity production in the Netherlands.

```
═══════════════════════════Weighted Summation══════════════════════════

-----------WEIGHTS---------               ----------RANKING--------
RANKING                 SCORES            RANK                SCORES
 1: Acidification        0.293             1: Gas               0.95
 2: Photochemical        0.193             2: Nuclear           0.80
 3: Greenhouse           0.143             3: Coal              0.06
 4: Eutrophication       0.110
 5: Export Acid          0.085
 6: Solid waste          0.065
 7: Landscape            0.048
 8: Land use             0.034
 9: Risk                 0.021
10: Nuclear waste        0.010




Select an option:
═Execute this multicriteria method═════════════════════════════C2.1═
F1.execute      F2.weights      F3.standardize  F4.save results F5.print result
F6.effects tab. F7.graph        F8.other MCA    F9.help         F10.continue
```

The left hand side of the screen shows the weights; the ranking and scores of the alternatives are shown on the right. Save this result but do not overwrite the previous result.

What to do:

Save the result:	**<F4>**
Select yes and assign information:	**Y YOURINFO <RETURN>**
Continue:	**<F10>**
Go to module D: **sensitivity analysis:**	**<F2>**

12.5. Sensitivity analysis

A central element in the energy discussion is the trade-off between effects related to emissions to air, such as the greenhouse effect, and effects related to nuclear power such as nuclear waste and risk. In this section the sensitivity of the rankings to changes in weights will be explored. The path this example takes through module B is emphasized in Figure 12.5.

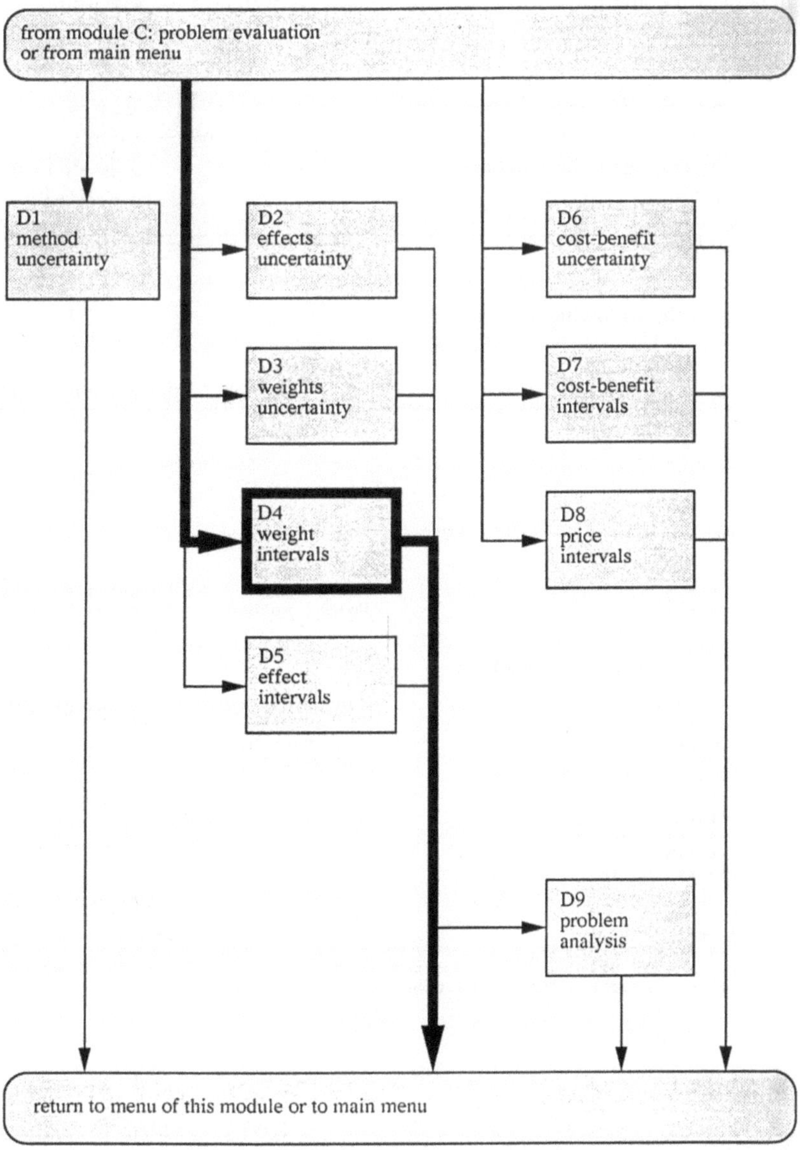

Figure 12.5. Sensitivity analysis. Electricity production in the Netherlands.

What to do:

Select **weight intervals**:	**<F4>**

Next you need to specify that you want to determine weight intervals and not weight combinations, the alternatives to be compared and the weight to be analyzed.

What to do:

Select **interval**:	**<F1>**
Select the alternatives to be compared:	**<F1>** and **<F2>**
Select the weight of **Risk**:	**<F9>**

This results in the following screen.

Screen 12.6. Weight interval. Electricity production in the Netherlands.

```
================================ WEIGHT INTERVAL ================================

        1 ------
          ------                         MCA method       : Weighted Summation
          ------                         Weight method    : Expected Value
          ------
          ------
          ------ Coal      > Nuclear
          ------
          ------                         The weight interval is calculated
          ------                         for effect        : Risk
          reverse   0.440                with rank number :    9
          ++++++                         and weight        :    0.021
          ++++++
          ++++++                         A rank reversal takes place
          ++++++                         at value :     0.440
          ++++++
          ++++++
    0.021 start  Nuclear   > Coal
        0 ++++++

Select an option:
==Save this interval=====================================================D4  ==
F1.save          F2.print         F3.information  F4.help          F5.other weight
F6.other alter.  F7.other MCA     F8.other anal.  F9.continue
```

For the weight of Risk this screen shows:

- the original value (start);
- the value where the ranking reverses (reversal);

- the interval where Nuclear is preferred to Coal (Nuclear > Coal);
- the interval where Coal is preferred to Nuclear (Coal > Nuclear).

It is shown that the weight assigned to Risk needs to be increased considerably to change the ranking order of Nuclear and Coal as calculated with the Weighted Summation method. In the same way the weight intervals related to other pairs of alternatives can be determined.

Only the relative weight of Risk was changed in calculating the weight intervals in Screen 12.6; the ratios of all other weights were held constant. If all weights are allowed to change freely, it becomes clear how sensitive the ranking is to overall changes of the weights.

What to do:

Select **other analysis**:	**<F8>**
Select **weight intervals**:	**<F4>**
Select **combination**:	**<F2>**
Select the **Nuclear** and **Gas** alternatives:	**<F1>** and **<F2>**

After some intermediate results this results in the following screen.

Screen 12.7. Weight combination. Electricity production in the Netherlands.

```
══════════════════════════ WEIGHT COMBINATION ══════════════════════════
                           (Weighted Summation)
The original weights                     ║Least distance reversal point
                                         ║(distance is:0.295)
-----------WEIGHTS-----------ALTERN--    ║-----------WEIGHTS-----------ALTERN--
RANKING               SCORES  RANK       ║RANKING               SCORES  RANK
 1:Acidification       0.293   1:Gas     ║ 1:Greenhouse          0.311   1:Nuclear
 2:Photochemical       0.193   2:Nuclear ║ 2:Solid waste         0.242     Gas
 3:Greenhouse          0.143   3:Coal    ║ 3:Acidification       0.166   3:Coal
 4:Eutrophication      0.110             ║ 4:Photochemical       0.109
 5:Export Acid         0.085             ║ 5:Eutrophication      0.062
 6:Solid waste         0.065             ║ 6:Export Acid         0.048
 7:Landscape           0.048             ║ 7:Landscape           0.027
 8:Land use            0.034             ║ 8:Land use            0.019
 9:Risk                0.021             ║ 9:Risk                0.012
10:Nuclear waste       0.010             ║10:Nuclear waste       0.006
                                         ║
Select an option:
══Save this combination═════════════════════════════════════════════D4  ══
F1.save         F2.print         F3.information  F4.help         F5.other alter.
F6.other MCA    F7.other anal.   F8.continue
```

Screen 12.7 shows that small changes in the weights result in a reversal of the ranking of alternatives Nuclear and Gas. You can repeat the analysis to compare the alternatives Nuclear and Coal using **<F5>**. This will show that for these alternatives the nearest weight combination that results in a rank reversal differs more from the original set of weights.

You have now completed the third example. You can either continue with one of the other examples or exit the program.

1. What to do to leave the program:

Select **continue**:	**<F8>**
Go to the **main menu**:	**<F3>**
Exit the program:	**<F7>**

You can now switch off your computer

2. What to do to continue to the next example:

Select **continue**:	**<F8>**
Go to the **main menu**:	**<F3>**
Select **utilities**:	**<F6>**
Change to a new session:	**<F1>**
Enter a new session name:	**NEW_NAME <RETURN>**
Answer yes to use the results of a previous session:	**Y**
Select the example you want to work with:	**<CURSOR><RETURN>**
Continue to the main menu:	**<F8>**

Continue to the chapter that describes the example you have selected.

13. SITING NUCLEAR PLANTS IN THE NETHERLANDS

13.1. Introduction

Compared with other European countries the share of nuclear power in total power production is at present very small in the Netherlands. After a long public debate in the early 1980s, in 1985 the Dutch government expressed the intention to build two new nuclear power plants with a capacity of 1000 MWe each. Given this decision the next question was where to locate them. After some initial scoping, nine potential locations for the plants were selected (Tweede Kamer, 1986[1]). These locations are shown in Figure 13.1.

This application includes four steps.

1. **Problem definition**
   In this step the effects table is created. The effects table includes the nine potential locations as alternatives and a number of criteria that reflect the suitability of these locations.

2. **Problem presentation**
   A graphic presentation provides an overview of the structure of the effects table.

3. **Problem evaluation**
   Two multicriteria methods: the Regime method and the Evamix method are applied to rank the alternatives.

4. **Sensitivity analyses**
   Method uncertainty is applied to analyze the relation between the rankings and the evaluation methods used to determine these rankings.

What to do:

Go to the DEFINITE directory:	**CD DEFINITE**	**<RETURN>**
Start the program:	**DEFINITE**	**<RETURN>**
Enter your own session name:	**YOURNAME**	**<RETURN>**
Answer yes to the question:	**Y**	
Select the example **Nuclear**:	**<CURSOR><CURSOR><CURSOR>**	**<RETURN>**

[1] Documents presented to Parliament for discussion.

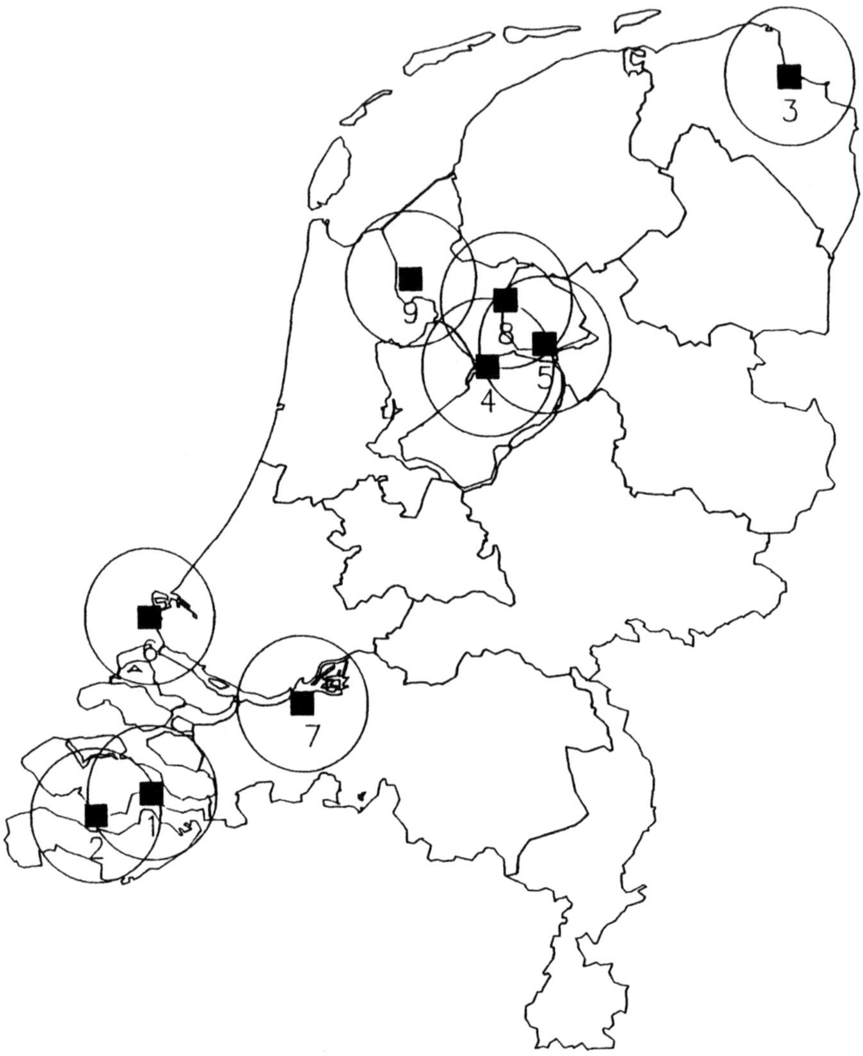

1. Bath/Hoedekenskerke
2. Borssele
3. Eems
4. Flevoland Noord
5. Ketelmeer
6. Maasvlakte
7. Moerdijk
8. Noord-Oost Polder
9. Wieringermeer

Figure 13.1. *Potential sites for nuclear plants in the Netherlands and the 20-km radius around these sites.*

It may take some time to go through the whole example. Return to the main menu and exit the program if you want to stop before reaching the end of the example. When you start the example again you can continue where you stopped.

13.2. Problem definition

Procedure A1 is used to create the effects table. The effects table includes the nine potential sites for a nuclear power station (the alternatives) and 15 criteria to evaluate these sites. The path this example takes through module A is emphasized in Figure 13.2.

What to do:

Enter module A: **problem definition:**	**<F1>**
Enter procedure A1: **create an effects table:**	**<F1>**

Information screens are presented upon entering each procedure. Hit **<ANY KEY>** to continue. The instruction to hit any key is not included in the "what to do: boxes".

This results in the following screen.

Screen 13.1. Effects table. Siting nuclear plants in the Netherlands.

effects table	Bath	Borssele	Eems	Flevo	Ketel	Maasvlak	Moerdijk
Population	-51.00	-49.00	-16.00	-27.00	-30.00	-43.00	-100.00
Evacuation	1	2	1	1	1	2	1
Agricult at risk	2	2	2	2	2	1	3
Industry at risk	1	4	3	2	1	5	3
Fr.water at risk	1	1	1	2	2	1	2
Cool-water quant	2	1	1	1	1	1	3
Cool-water qual.	2	1	2	3	3	1	2
Air pollution	2	2	2	2	2	1	2
Thermal poll.	3	2	2	2	3	1	2
Indirect landuse	2	3	3	2	1	2	4
Landscape	3	1	1	1	3	1	2
Nat. environment	3	1	3	1	2	1	1
National grid	2	2	3	1	1	2	2
Infrastructure	2	1	1	2	2	1	1
Coal-location	3	6	4	3	2	7	5

```
Select an option:
=Add an alternative=====================================================A1  =
F1.alternative   F2.effect        F3.descriptor   F4.edit        F5.units
F6.time          F7.information   F8.help         F9.continue    F10.more
```

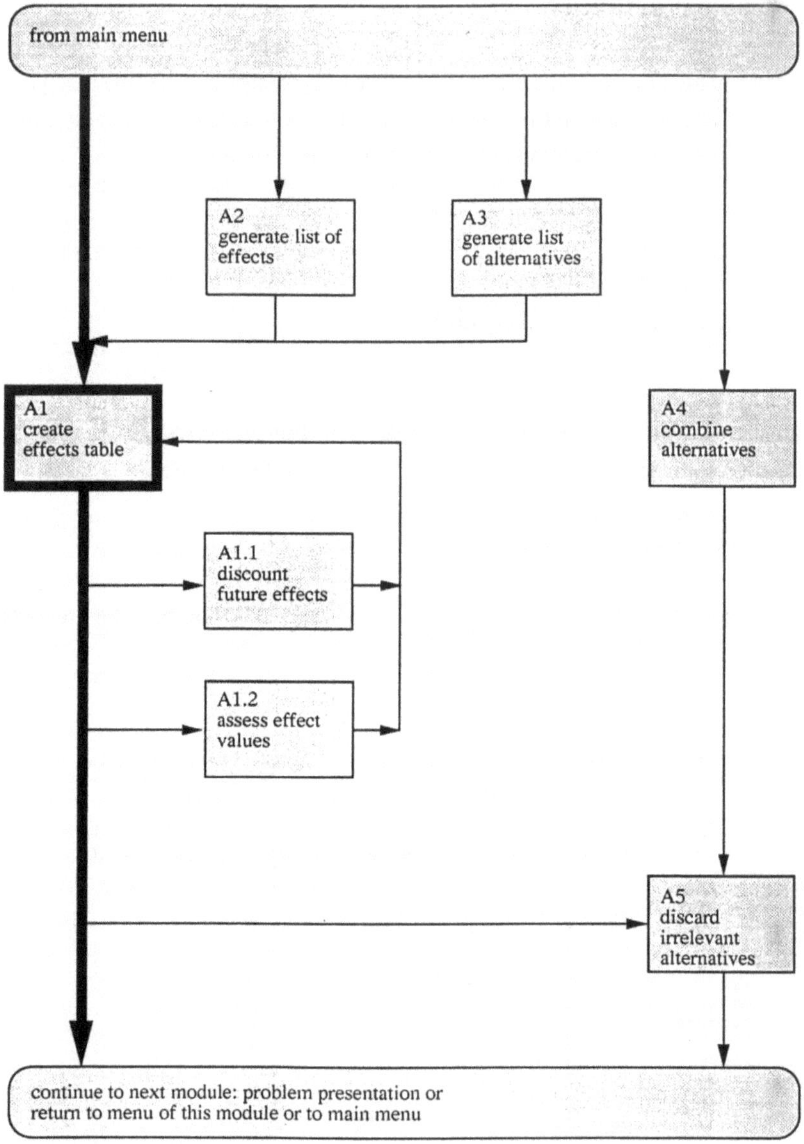

Figure 13.2. Problem definition. Siting nuclear plants in the Netherlands.

The full names of the evaluation criteria are listed below:

- Population
- Evacuation
- Agriculture at risk
- Industry at risk
- Fresh water at risk
- Cool-water quantity
- Cool-water quality
- Air pollution
- Thermal pollution
- Indirect land use
- Landscape
- Natural environment
- National grid
- Infrastructure
- Coal-location

The effects table of this evaluation problem is shown in Screen 13.1, where nine potential locations are scored according to 15 criteria. Only the score for population around a site is measured on a quantitative scale. All other scores are measured on a qualitative scale: a score 1 is assigned to the best alternative, 2 to the second best, etc. (see the appendix at the end of this chapter for a definition of the criteria). The qualitative scores are the result of expert judgement of a panel of experts. This type of assessment could be supported using procedure A1.2: assess scores.

The effects table is the final result of problem definition. You can now continue to the next module: problem presentation.

13.3. Problem presentation

Graphic presentation is used in this example to analyze the structure of the effects table. The path this example takes through module B is emphasized in Figure 13.3.

What to do:

Continue:	**<F9>**
Go to **module B**:	**<F2>**
Select **graphic presentation**:	**<F3>**
Rank the alternatives:	**<F4>**

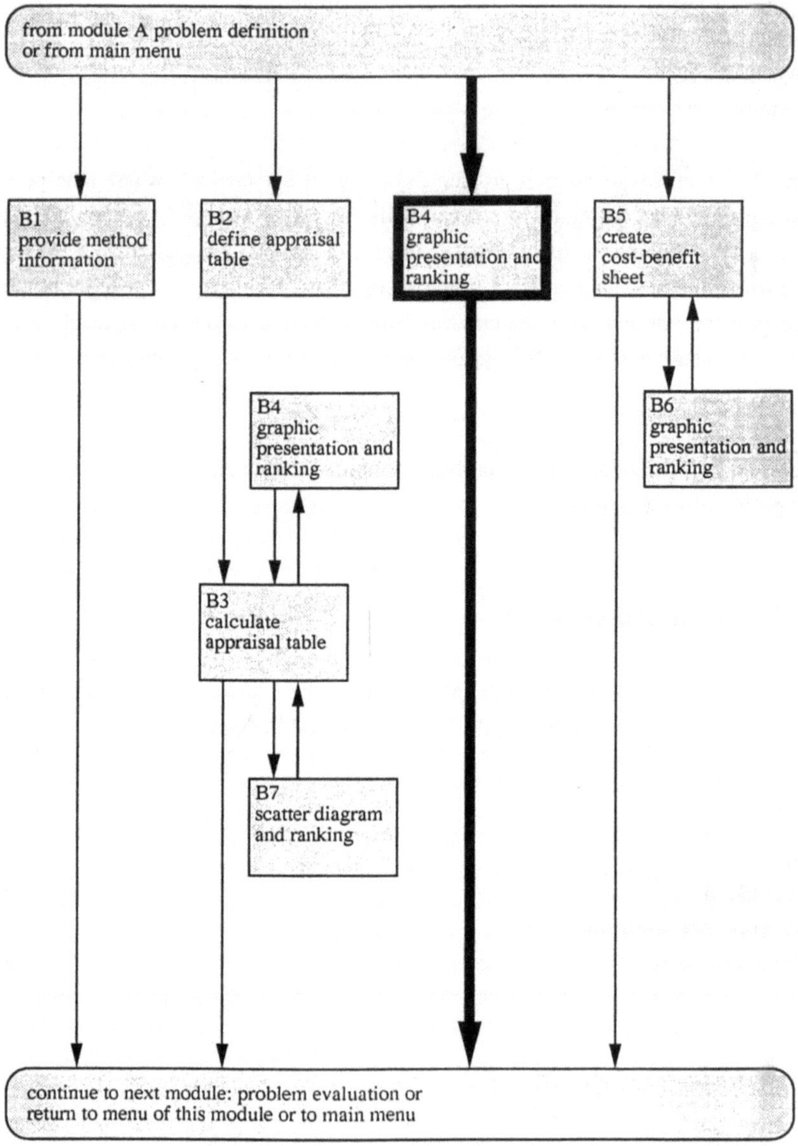

Figure 13.3. Problem presentation. Siting nuclear plants in the Netherlands.

This results in the following screen.

Screen 13.2. Graphical presentation. Siting nuclear plants in the Netherlands.

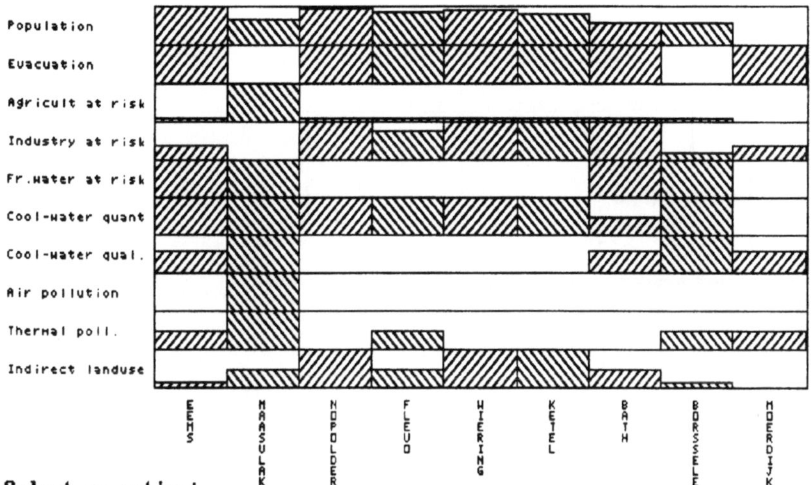

Select an option:
=Rank the alternatives ═══════════════════════
F1.exchange row F2.exchange clm F3.print F4.rank **F5.information**
F6.Help F7.continue

The graph clearly show the large number of equal scores (ties) for certain criteria such as Evacuation or Agriculture at risk. The graph also shows the similarities between the scores of certain alternatives such as NO Polder and Wieringermeer.

13.4. Problem evaluation

As a first step you could consult method information for advice on the evaluation method to be applied. Because in this example most of the effects are measured on a qualitative scale, the Regime and Evamix methods are used to rank the alternatives. Procedure C3 is used to create weights. The path this example takes through module C is emphasized in Figure 13.4.

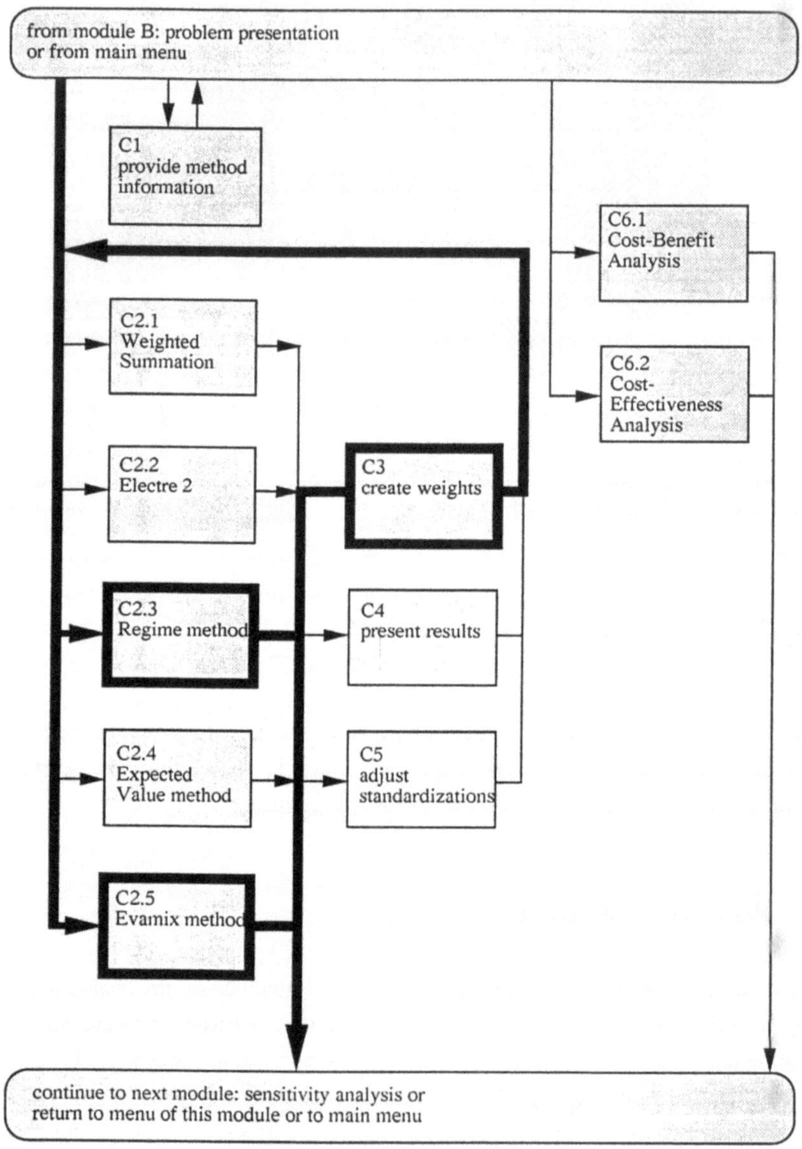

Figure 13.4. Problem evaluation. Siting nuclear plants in the Netherlands.

What to do:

Continue:	<F7>
Go to **module C**:	<F5>
Select **multicriteria analysis**:	<F1>
Select **method information**:	<F1>
Stop and continue with the program:	<F3>
Select the **Regime method**:	<F4>

You have now entered the Regime method. Before you can start calculation you need to assign weights to the criteria. Since the Regime method requires only ordinal weights it is sufficient to specify a priority order of the criteria. The experts involved in the analysis defined four different priority orders: a standard priority order, an order with priority on economic criteria, an order with priority on environmental criteria and an order with priority on land use criteria.

In this example the standard priority is used as shown below. You can now enter this order or retrieve it from disk.

1. Population
2. Industry at risk
3. Agriculture at risk
Fresh water at risk
5. Cool-water quantity
Cool-water quality
Thermal pollution
Coal location
9. Air pollution
Landscape
Natural environment
National grid
Infrastructure
14. Evacuation
Indirect land use

What to do:

Select **weights**:	<F2>
Select **rank order**:	<F1>
Retrieve the weights from disk:	<F2>
Execute:	<F1>

After some time (DEFINITE has to calculate the ranking many times), the following intermediate result is produced.

Screen 13.3. Intermediate results of the Regime method.

Siting nuclear plants in the Netherlands.

Probabilities	Bath	Borssele	Eems	Flevo	Ketel	Maasvlak	Moerdijk
Bath		0.19	0.06	0.18	0.03	0.02	0.99
Borssele	0.81		0.02	0.06	0.45	0.26	0.99
Eems	0.94	0.98		0.56	0.87	0.51	1.00
Flevo	0.82	0.94	0.44		0.43	0.37	1.00
Ketel	0.97	0.55	0.13	0.57		0.19	0.99
Maasvlak	0.98	0.74	0.49	0.63	0.81		1.00
Moerdijk					0.01		
NOPolder	0.99	0.75	0.33	0.87	1.00	0.37	1.00
Wiering	0.93	0.57	0.17	0.85	0.99	0.24	0.98

```
Examine the table using the cursor keys
==End this procedure and continue=========================================C2.3==
F1.continue      F2.print
```

This intermediate result shows the results of all pairwise comparisons of the alternatives. Due to the use of a random generator small variations will occur each time this table is calculated. The scores in this table indicate the probability that given the ordinal information on scores the row alternative ranks over the column alternative. For example, the value of 0.81 in the comparison between Borssele and Bath indicates that, given the uncertainty that results from the ordinal information, Borssele ranks over Bath with 81% certainty. Hit **continue <F1>** to produce the final result.

```
═══════════════════════════════Regime Method═══════════════════════════════

-----------WEIGHTS---------                ----------RANKING--------
RANKING                                    RANK               SCORES
 1: Population                              1: Eems             0.80
 2: Industry at risk                           NOPolder         0.79
 3: Agricult at risk                        3: Maasvlak         0.75
    Fr.water at risk                        4: Wiering          0.59
 5: Cool-water quant                        5: Flevo            0.53
    Cool-water qual.                        6: Ketel            0.43
    Thermal poll.                           7: Borssele         0.41
    Coal-location                           8: Bath             0.20
 9: Air pollution                           9: Moerdijk
    Landscape
    Nat. environment
    National grid
    Infrastructure
14: Evacuation
    Indirect landuse

Select an option:
═Execute this multicriteria method═════════════════════════════════C2.3═
F1.execute       F2.weights       F3.standardize  F4.save results F5.print result
F6.effects tab.  F7.graph         F8.other MCA    F9.help         F10.continue
```

The left hand side of the screen shows the ordinal weights; the ranking and scores of the alternatives are shown on the right. To check whether another method will produce the same result the ranking is determined again using the Evamix method.

What to do:

Save this result:	**<F4>**
Do **not** change information:	**N**
Go to **other MCA**:	**<F8>**
Select the **Evamix method**:	**<F6>**

You have now entered the screen of the Evamix method. Since the Evamix method requires quantitative weights you will need to determine the weights again.

What to do:

Select **weights**:	**<F2>**
Select **Expected Value method**:	**<F1>**
Retrieve the weights from disk:	**<F2><ESC>**
(Select **wexp.001** if a choice is offered)	
Execute:	**<F1>**

This results eventually in the following screen.

Screen 13.5. Results Evamix method. Siting nuclear plants in the Netherlands.

```
═══════════════════════════════════Evamix Method═══════════════════════════════════

-----------WEIGHTS---------                  ----------RANKING--------
RANKING                  SCORES              RANK                 SCORES
 1: Population            0.221               1: Maasvlak          0.11
 2: Industry at risk      0.155               2: Eems              0.05
 3: Agricult at risk      0.110               3: NOPolder          0.03
    Fr.water at risk      0.110                  Bath              0.02
 5: Cool-water quant      0.064                  Wiering           0.01
    Cool-water qual.      0.064                  Borssele          0.01
    Thermal poll.         0.064                  Ketel
    Coal-location         0.064                  Flevo
 9: Air pollution         0.027               9: Moerdijk         -0.24
    Landscape             0.027
    Nat. environment      0.027
    National grid         0.027
    Infrastructure        0.027
14: Evacuation            0.007
    Indirect landuse      0.007

Select an option:
═Execute this multicriteria method═════════════════════════════════════════C2.5═
F1.execute       F2.weights       F3.standardize  F4.save results F5.print result
F6.effects tab.  F7.graph         F8.other MCA    F9.help         F10.continue
```

The ranking determined with the Evamix method differs considerably form the ranking determined with the Regime method. Although both methods point Moerdijk out as the worst alternative and include Eems, Maasvlakte and NO Polder within the first three there are large differences in the remainder of the ranking. The Evamix method, for example, selects Maasvlakte as the best alternative while this alternative is ranked second using the Regime method. The sensitivity of the results to the method used results from the small differences between the alternatives. This is analyzed further in the next section.

13.5. Sensitivity Analysis

The method uncertainty procedure is used to analyze the relation between the ranking and the evaluation method that is used to determine this ranking. The path this example takes through module D is emphasized in Figure 13.5.

What to do:

Save this result:	<F4>
Do not change information:	**N**
Continue:	<F10>
Go to module D:	<F2>
Select **method uncertainty**:	<F1>
Calculate:	<F1>
Select **line graph**:	<F2>
Specify 5 alternatives to be compared:	**5**
Select **Borssele, Eems, Maasvlak, Moerdijk, NO Polder**:	<F2><F3><F6><F7><F8><RETURN>

This results in the following screen.

Screen 13.6. Method uncertainty. Siting nuclear plants in the Netherlands.

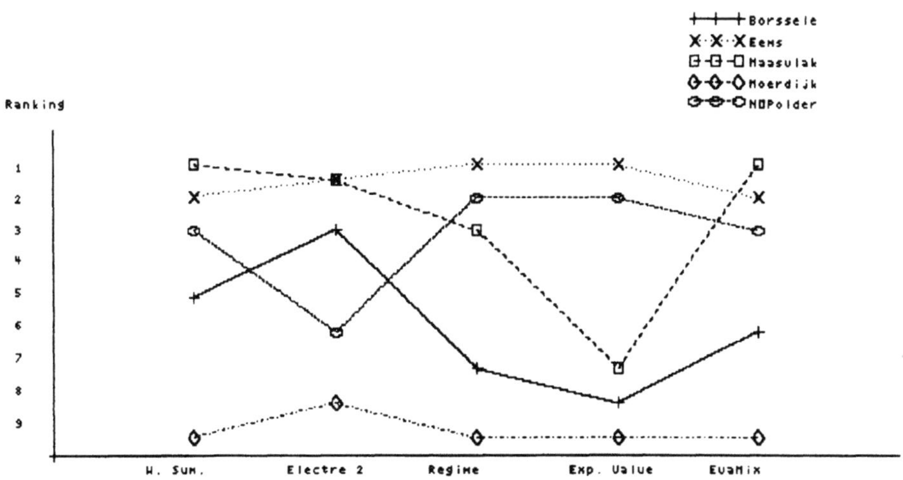

Select an option:
=Print the diagram
F1.print F2.continue

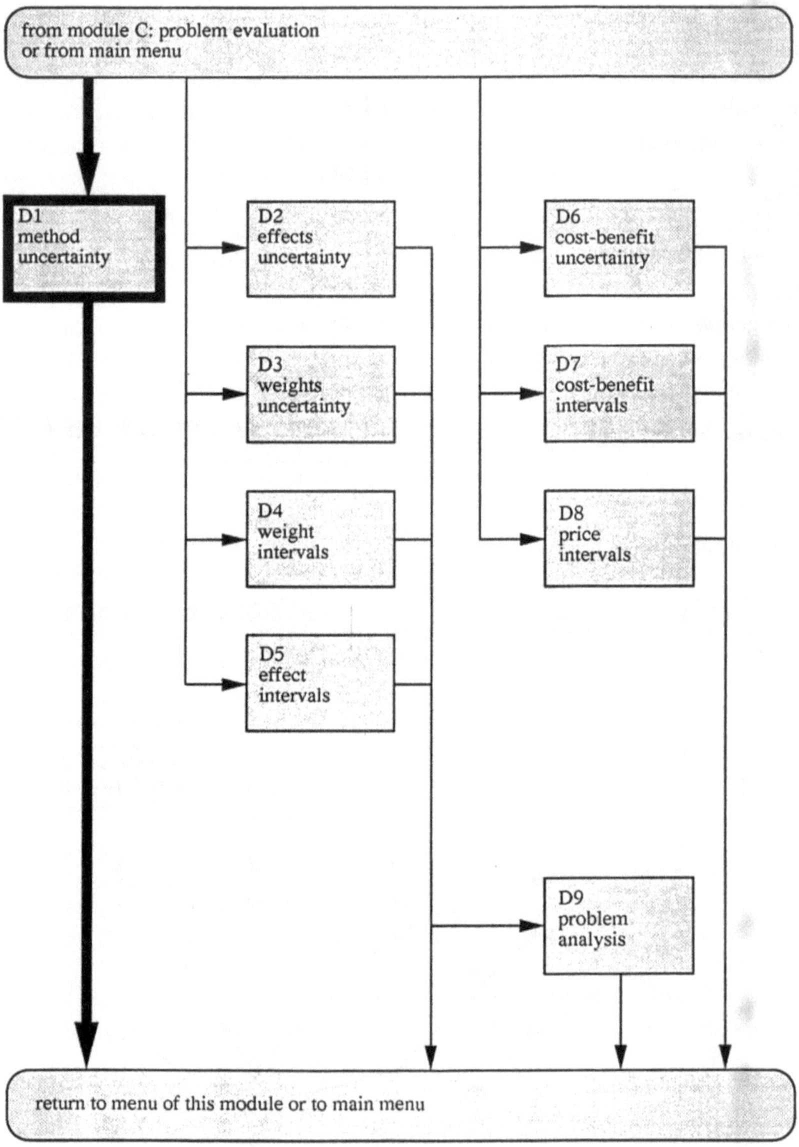

Figure 13.5. Sensitivity analysis. Siting nuclear plants in the Netherlands.

This diagram shows the rankings according to all five multicriteria methods available in DEFINITE. It is clear that Moerdijk ranks last for all methods. The conclusion that Moerdijk is the least preferred of these five alternatives is therefore not subject to method uncertainty. The same holds for the conclusion that Eems is better than NO Polder. The ranking of Maasvlakte, Eems and NO Polder, however, is strongly related to the method applied.

What to do:

Leave the diagram:	**<F2>**
Continue:	**<F9>**
Go to the **main menu**:	**<F2>**
Exit the program:	**<F7>**

Appendix: Definition of criteria used

Population	A weighted sum of population around a location was calculated to quantify this score. The weight assigned decreases with distance. The result is standardized by dividing it by the maximum score. A minus sign is added to indicate that the criterion is a cost criterion.
Evacuation	This score reflects the availability of sufficient transport infrastructure.
Agriculture at risk	This score reflects the location of agricultural land in the vicinity.
Industry at risk	This score reflects the size and importance of industry near the location.
Fresh water at risk	This score reflects the quantity of fresh water that may be affected by a nuclear plant at each location.
Cool-water quantity	This score represents the quantity of water available for cooling the nuclear plant.
Cool-water quality	This score represents the capacity of the coolant to flush out pollution originating from a nuclear plant at each location.
Air pollution	It is assumed that the nuclear plant is an alternative to a conventional coal power plant; this implies that the best score is given to the most polluted location.
Thermal pollution	The score reflects the availability to users of the waste heat generated by the plant.
Indirect land use	This score reflects limitations on potential land use around a nuclear plant.
Landscape	This score reflects the visual effects on the landscape and the extent to which a nuclear plant fits in with existing activities.
Natural environment	This score reflects expected damage to the natural environment.
National grid	This score reflects the proximity of high-voltage lines and connector stations.
Infrastructure	This score reflects the availability of transport and other infrastructure near the site.
Coal-location	It is assumed that the nuclear plant is an alternative to a conventional coal power plant. The score reflects the cost of the lost opportunity to build a coal plant at the site if a nuclear plant is constructed.

APPENDIX. TECHNICAL MANUAL

1. Introduction

You do not need this technical manual to work with the program. The technical manual is only included to support installation in case of unexpected problems and for those who are interested in the structure of the program.

2. Equipment you need

To be able to run DEFINITE you need the following equipment:
- an IBM personal computer or 100% compatible;
- a 3½ inch 720 KB diskdrive;
- a hard disk with at least 2.5 MB available;
- 640 KB memory with at least 570 KB available;
- a CGA, EGA, VGA, or compatible graphics card;
- MS-DOS 2.0 or higher.

Memory
DEFINITE uses 570 KB memory. If you have resident programs on your computer that require a lot of memory space it is better to deactivate them while running DEFINITE. If you do not know how much memory you have available just run the program. DEFINITE provides a message if the necessary memory is not available. Free memory and available space on the hard disk is reported by the MS-DOS **CHKDSK** or **MEM** command.

Most modules of the program will still be executed if less than 570 KB is available. If you select a procedure that requires more memory space than available the program provides a warning. The program then returns to the main menu.

Networks
DEFINITE is not designed for use on a network. If you run DEFINITE on a network the memory space allocated to the network software will prevent you from using some of the available procedures. Some networks include special memory allocation software that may prevent DEFINITE from starting up. It is not possible to print directly through a network. If you run DEFINITE on a network you are advised to generate a report in module E, save your results as an ASCII file (option 2) and process the file in a word processor before printing. Graphic screens can be printed by using a screen save program such as the GRAB program of WordPerfect, and printing the result using your word processor.

Graphics card

The install program automatically installs DEFINITE for use with a EGA or VGA graphics card. Insert disk 'CGA' and type: **A: CGA<RETURN>** if you want to install DEFINITE for a CGA card. Other graphics cards are not supported. If your graphics card is not supported or if you have installed DEFINITE incorrectly the program will provide you with a warning when you enter a procedure that requires the graphics card and you will return to the main menu. You will still be able to run all procedures that do not require a graphics card.

5¼" 360KB disks copy utility

A special 3½ inch to 5¼ inch copy utility is available to accommodate for users who do not have a 3½ inch diskdrive but have access to a computer with both types of disk drives.

1. Format seven 5¼ inch (360 KB) disks
2. Label these disks disk '1:7', disk '2:7'...disk '7:7'
3. Insert the DEFINITE 3½ inch disk '1:3' in drive A
4. Insert your 5¼ inch disk '1:7' in drive B
5. Change to drive A: **A:<RETURN>**
6. Start the copy utility: **COPY360K<RETURN>**
7. Follow the instructions of the copy utility
8. Install the program as described under installation
   (Pay careful attention to the requested disk numbers while performing steps 7 and 8)

How to copy the DEFINITE CGA disk

1. Format three 5¼ inch (360 KB) disks
2. Label these disks 'CGA 1:3', 'CGA 2:3' and 'CGA 3:3'
3. Insert the DEFINITE 3½ inch disk 'CGA' in drive A
4. Insert your 5¼ inch disk 'CGA 1:3' in drive B
5. Change to drive A: **A:<RETURN>**
6. Start the CGA copy utility: **CGA360K<RETURN>**
7. Follow the instructions of the copy utility
8. Install the CGA part as described under graphics card

A=B and B=A

In case your drives are reversed (the B drive is the 3½ inch drive and the A drive the 5¼ inch drive), you need to temporarily reverse the definitions of your drives.

1. Change to drive C: **C:<RETURN>**
2. Change the drive definitions: **ASSIGN A=B;B=A<RETURN>**
3. Change to drive A: **A:<RETURN>**
   (Note that after ASSIGN your drives are reversed. Read A for B and B for A while following the instructions of the copy utility.)

(This utility also creates the necessary install files by renaming the files INST360D.BAT and BOS360D.BAT to INSTALL.BAT and BOSINST.BAT)

Printing

DEFINITE supports EPSON compatible printers that are directly linked to the personal computer DEFINITE has been installed on. If you want to print your results on a different type of printer we recommend that you save all results you want to print within each procedure. Then go to module E: report generation, select option 2 to save your results as an ASCII file and process the file in a word processor before printing.

The report generated does not include results of any of the graphic procedures. These results can be printed directly from a graphic procedure if an EPSON compatible printer is available. If this is not the case graphic screens should be saved using a screen save program, such as the GRAB program of WordPerfect, and then printing these screens using the appropriate word processor.

3. Installing DEFINITE

To be able to run DEFINITE from the hard disk you have to do the following to install the program on hard disk.

1. Make sure the DEFINITE disks are in the read only position;
2. Insert the DEFINITE disk '1:3' into the diskdrive;
3. Create a DEFINITE directory on the hard disk: **MD DEFINITE <RETURN>**;
4. Change to the DEFINITE directory: **CD DEFINITE <RETURN>**;
5. Start the installation program: **A:INSTALL <RETURN>**;
6. Follow the instructions of the installation program.

You can now start the program: **DEFINITE <RETURN>**

If installation fails due to insufficient disk space, create more space and try again. If automatic installation does not work you can also install DEFINITE by copying the files across to the appropriate directories. Refer to the next two sections for a list of directories to be created and for a list of files to be copied to the various directories.

4. DEFINITE structure

DEFINITE includes a number of programs that are linked to the various modules and a central program BOSDA.EXE that organizes communication between programs. Only the central program is always present in memory.

DEFINITE includes the following programs.

BOSDA.EXE	This is the central program. This program includes the main menu and organizes the loading of all other modules.
TITEL.EXE	Includes the title page and utilities.
BLOKA.EXE	All procedures of module A.
BLOKB.EXE	All procedures of module B.
BLOKC.EXE	All procedures related to multicriteria analysis included in module C.
BLOKCKB.EXE	All procedures related to Cost-Benefit analysis included in module C.
BLOKD.EXE	All procedures of module D except procedure D1.
GRAFIEK.EXE	All routines that support graphic screens.
MAAKRPT.EXE	Routines to generate the files that contain the chapters of the report.
BLOKE.EXE	All routines to support word processing within module E.

5. The DEFINITE directories

The install program generates a number of directories on hard disk and copies all files to their appropriate directory. This section includes a description of these directories and the files included in each of these directories.

DEFINITE directory
This is the main directory for the program. It includes the files:

*.fon	character files.
definite.bat	the batch file to start the program (execute\bosda).
bosinst.bat	the batch file that installed the program.
error.msg	a file that includes the Pascal error messages.

DEFINITE\EXECUTE directory
This directory includes the program itself. This directory includes all *.EXE files as described in the previous section.

DEFINITE\INFORM directory

This directory includes only the following text files:

*.inf	Includes one page of text that is displayed upon entering each procedure.
*.hlp	Includes the text of the help screens that are available for each procedure.
category.*	The list of effects available in procedure A2.
bloke.mac	The macros of the word processor in module E.

DEFINITE\SESSIES\ directory

Each problem is stored in a separate session. The program uses MS-DOS commands to create a directory for each session specified by the user.

The program includes five example sessions. These example sessions are stored in the following directories.

DEFINITE\SESSIES\EXAMPLE
DEFINITE\SESSIES\FOREST
DEFINITE\SESSIES\SOIL
DEFINITE\SESSIES\ENERGY
DEFINITE\SESSIES\NUCLEAR

A session always includes the files:

Inhoud	This file contains all file names and information lines of a session.
Heenweer	This file supports the main program in switching between the modules.

The files that can be created in a session are listed below.

ALTINFO	.001 - a text file linked to the effects table (A1)
EFFETAB	.001 - effects table (A1)
TIME	.00n - time-effects tables (A1.1)
EFFECTS	.001 - list of effects (A2)
ALTS	.001 - list of alternatives (A3)
COMBINED	.001 - combined alternatives (A4)
CONSTR	.001 - list of constraints (A5)
DEFIN	.001 - definitions table (B2)
EVAL	.001 - appraisal table (B3)
CBSHEET	.001 - cost-benefit table (B5)
WEXP	.00n - weight-vector Expected Value method (C3)
WDIRECT	.00n - weight-vector direct assessment (C3)
WPAIR	.00n - weight-vector Pairwise comparison (C3)
WORD	.00n - weight-vector ordinal (C3)
SUMM	.00n - MCA-result Weighted Summation (C2.1)
ELECTRE	.00n - MCA-result Electre (C2.2)

EXP	.00n - MCA-result Expected Value method (C2.3)
EVAMIX	.00n - MCA-result Evamix method (C2.4)
REGIME	.00n - MCA-result Regime method (C2.5)
STAND	.00n - standardization vector
CBA	.001 - result Cost-Benefit analysis (C6.1)
CEA	.001 - result Cost-Effectiveness analysis (C6.2)
MUNC	.00n - result method uncertainty (D1)
SCOREUNC	.00n - result effects uncertainty (D2)
WUNC	.00n - result weights uncertainty (D3)
WINT	.00n - result weight interval (D4)
WCOMB	.00n - result weight combination (D4)
SCOREINT	.00n - result effect interval (D5)
CBAUNC	.001 - result cost-benefit uncertainty (D6)
CBAINT	.00n - result cost-benefit interval (D6)
PRICEINT	.00n - result price interval (D8)
PRANALY	.001 - result problem analysis (D9)
GRAPH	.001 - table to be printed (B4, C4).

6. Remove DEFINITE from your harddisk

1. Select **utilities: <F6>**
2. Remove all user sessions: **<F2>**
3. Name the last session last: **LAST<RETURN>**
4. Return to the main menu and exit the program: **<F8><F7>**
5. Run the special delete program: **DELDEF<RETURN>**

7. Remarks

- **Installation on a network**
  Installation on a network is not recommended. Joint use of more than one user may cause difficulties. The memory space used by the network itself may not leave DEFINITE sufficient memory. Memory allocation programs installed on some networks may prevent DEFINITE from being started.

- **Printing on other types of printers**
  DEFINITE only supports EPSON compatible printers. To print text or figures on, for example, a Postscript printer you need to use your word processor as an intermediary. Refer to the manual of your word processor to import and print text and figures.

- **Personal computers under OS/2**

  DEFINITE cannot be used in combination with OS/2. Entering MS-DOS from OS/2 followed by starting DEFINITE may also cause problems.

- **Hard disk full while using DEFINITE**

  Definite creates a directory for each session. Use utilities to remove redundant sessions.

- **WINDOWS**

  DEFINITE can be run from WINDOWS. It should be noted, however, that certain settings within WINDOWS, for example settings related to available RAM and keyboard, may prevent DEFINITE from performing all functions.

MIX
Papier aus verantwortungsvollen Quellen
Paper from responsible sources
FSC® C105338

If you have any concerns about our products,
you can contact us on
ProductSafety@springernature.com

In case Publisher is established outside the EU,
the EU authorized representative is:
Springer Nature Customer Service Center GmbH
Europaplatz 3, 69115 Heidelberg, Germany

Printed by Libri Plureos GmbH
in Hamburg, Germany